T0135797

Increasing the delivery pressure of high-pressure gasoline pumps for direct injection engines by the use of ceramic components

Zur Erlangung des akademischen Grades

Doktor der Ingenieurwissenschaften

der Fakultät für Maschinenbau

Karlsruher Institut für Technologie (KIT)

genehmigte
Dissertation
von

Dipl.-Ing. Christophe Pfister

Tag der mündlichen Prüfung:	26.06.2013
Hauptreferent:	Prof. Dr.-Ing. Ulrich Spicher
Korreferent:	Prof. Dr. rer. nat. M. J. Hoffmann

Forschungsberichte aus dem
Institut für Kolbenmaschinen
Karlsruher Institut für Technologie
Hrsg.: Prof. Dr.-Ing. U. Spicher

Bibliografische Information der Deutschen Nationalbibliothek

Die Deutsche Nationalbibliothek verzeichnet diese Publikation in der
Deutschen Nationalbibliografie; detaillierte bibliografische Daten sind
im Internet über http://dnb.d-nb.de abrufbar.

ISBN 978-3-8325-3482-0
ISSN 1615-2980

Logos Verlag Berlin GmbH
Comeniushof, Gubener Str. 47,
10243 Berlin
Tel.: +49 030 42 85 10 90
Fax.: +49 030 42 85 10 92
INTERNET: http://www.logos-verlag.de

Vorwort des Herausgebers

Der moderne Verbrennungsmotor stellt aufgrund seiner von keiner anderen Technologie erreichten Kombination aus Leistungsdichte und Alltagstauglichkeit bei gleichzeitig moderaten Produktionskosten die wichtigste Antriebsquelle im Straßenverkehr dar. Der heutige Vorsprung, wie auch das erhebliche Weiterentwicklungspotenzial, trägt dazu bei, dass in Konkurrenz zu heute noch in der Forschung befindlichen alternativen Antriebskonzepten die dominierende Rolle des Verbrennungsmotors auf absehbare Zeit wohl kaum angefochten werden wird. Die teilweise noch in der Entwicklung befindlichen, teilweise jedoch auch bereits in die Serienproduktion einfließenden Innovationen machen deutlich, dass trotz bereits jahrzehntelanger Weiterentwicklung noch erhebliche Potenziale in dieser Antriebstechnik schlummern.

Die vielfältigen Anforderungen an den verbrennungsmotorischen Antrieb, wie Abgaslimitierung und die Forderung nach weiterer Verbrauchsreduktion, aber auch Komfort- und Leistungsansprüche, stehen häufig im Widerspruch zueinander. Auch aufgrund der hohen Komplexität moderner Motoren und der im Motor ablaufenden Prozesse steigt der Aufwand für Forschung und Entwicklung immer weiter an. Die erforderlichen Innovationen können durch einen verstärkten Transfer wissenschaftlicher Ergebnisse in die Praxis beschleunigt werden.

Der Austausch von Forschungsergebnissen und Erfahrungen zwischen Hochschulen und der Industrie ist deshalb von großer Bedeutung für die Weiterentwicklung und Optimierung von Motoren. Neben unseren Veröffentlichungen auf internationalen Tagungen und in Fachzeitschriften versuchen wir mit vorliegender Schriftreihe aktuelle Forschungsergebnisse des Instituts für Kolbenmaschinen des Karlsruher Instituts für Technologie (KIT) den Fachkollegen aus Wissenschaft und Industrie zugänglich zu machen.

Der Schwerpunkt unserer Aktivitäten liegt in der Optimierung der motorischen Brennverfahren mit der besonderen Zielsetzung einer Senkung von Kraftstoffverbrauch und Schadstoffemissionen. Zur Lösung dieser Aufgaben stehen an unseren Motorenprüfständen neue und technisch hoch entwickelte Messtechniken zur Verfügung.

In zahlreichen nationalen und internationalen Forschungsvorhaben versuchen wir den Kenntnisstand über Teilprozesse wie Ladungswechsel, Gemischbildung, Verbrennung und Schadstoffentstehung zu erweitern, um wichtige Grundlagen für die Entwicklung zukunftsorientierter Motorkonzepte bereitzustellen. Vielfältige Industrieprojekte mit interessanten Aufgabenstellungen ermöglichen den Ausbau des Kenntnisstandes über motorische Prozesse, die weitere Verbesserung unserer Messtechniken sowie die Erweiterung unserer Prüfstandseinrichtungen.

Die Ausbildung von Studenten im Bereich der Motorenforschung - ob im Rahmen von Studien- und Diplomarbeiten oder als studentische Hilfskräfte - sichert wiederum den Nachwuchs an Fachkräften für Fahrzeug-, Motoren- und Zulieferindustrie im In- und Ausland.

Im vorliegenden Band 4/2013 berichtet Herr Pfister über experimentelle Untersuchungen zur Realisierung von Förderdrücken bis 800 bar durch die Nutzung von keramischen Komponenten in einer Hochdruckpumpe für die Direkteinspritzung von Benzin. Dieses Druckniveau ermöglicht eine schnellere Gemischbildung in Verbrennungsmotoren bei direkter Kraftstoffeinspritzung, und damit eine bessere Ausnutzung des Potenzials dieser Technologie. Neben grundlegenden tribologischen Untersuchungen von Reibungszahlen mit verschiedenen Materialpaarungen in den Gleitsystemen werden Ergebnisse von Effizienzmessungen vorgestellt. Die Analyse der beanspruchten Oberflächen ist von besonderem Interesse, in dem sie wichtige Informationen über die Reibungs- und Verschleißmechanismen liefert. Zyklische Schwankungen und Lärmemissionen mit verschiedenen Werkstoffpaarungen werden durch eine Spektralanalyse der Verläufe von Antriebsmomenten charakterisiert. Die Ergebnisse aus der vorliegenden Arbeit zeigen, dass die Nutzung ingenieurkeramischer Werkstoffe in den Gleitsystemen der Benzin-Hochdruckpumpe eine deutliche Steigerung des Förderdrucks erlaubt. Dadurch können auch die Vorteile der Benzin-Direkteinspritzung besser ausgenutzt werden, da höhere Einspritzdrücke zu signifikant geringeren Schadstoffemissionen führen. Dies konnte bereits in Untersuchungen zur Benzin-Direkteinspritzung mit Einspritzdrücken im Bereich bis zu 1000 bar mit Reduzierungen von bis zu 99 % bei den Rohemissionen von Rußpartikeln nachgewiesen werden.

Karlsruhe, Juli 2013   Prof. Dr.-Ing. Ulrich Spicher

Zusammenfassung

Die strahlgeführte Benzin-Direkteinspritzung erlaubt es, den Kraftstoffverbrauch von Verbrennungsmotoren im Vergleich zur Saugrohr- oder homogenen Direkteinspritzung zu reduzieren. Jedoch führt diese Technologie zu neuen Herausforderungen. Da der Kraftstoff direkt in den Brennraum eingespritzt wird, ist die Zeitspanne zur Verdampfung und Gemischbildung besonders kurz. Falls noch flüssiger Kraftstoff während der Verbrennung vorliegt, bildet sich Ruß, was zu deutlich erhöhten Partikelemissionen führt. Die Verdampfung des flüssigen Kraftstoffes kann durch eine Einspritzdruckerhöhung signifikant beschleunigt werden.

Konventionelle Werkstoffe, die in Benzin-Hochdruckpumpen zum Einsatz kommen, versagen bei über 200 bar Förderdruck, da die Schmierfähigkeit des Ottokraftstoffs zu niedrig ist. Diese Schwierigkeiten können durch den Einsatz anderer Werkstoffe überwunden werden. Ingenieur-Keramiken weisen eine hohe Härte und gute chemische Beständigkeit auf und können helfen, den Förderdruck von Hochdruckpumpen deutlich zu erhöhen.

Die in diesem Dokument präsentierten Ergebnisse haben das Ziel, die Eignung von keramischen Komponenten in den hochbeanspruchten Gleitsystemen einer Hochdruckpumpe zu bewerten. Dazu wurden zwei Versuchsträger, die im Rahmen des Sonderforschungsbereichs 483 („Hochbeanspruchte Gleit- und Friktionssysteme auf Basis ingenieurkeramischer Werkstoffe") entwickelt wurden, verwendet. Eine Einkolbenpumpe mit Kraftsensoren ermöglichte, die Reibungszahlen im Nocken/Gleitschuh-System und im Kolben/Zylinder-System bis zu 500 bar Förderdruck zu messen. Eine serien-nähere Dreikolbenpumpe, die für 800 bar Förderdruck ausgelegt ist, erlaubte es, die globalen Reibungsverluste in der Pumpe über deren mechanischen Wirkungsgrad zu berechnen. Der Fokus der Untersuchungen liegt auf dem Nocken/Gleitschuh-System, in dem die höchsten Kontakt- und Reibungskräfte auftreten. Untersucht wurden Siliziumkarbid, Sialon und Siliziumnitrid, da diese Werkstoffe einen hohen Verschleißwiderstand, niedrige Reibungskoeffizienten und eine gute chemische Beständigkeit mit Ottokraftstoff oder Ethanol aufweisen. Nocken aus Siliziumkarbid wurden mit einer Mikrotextur versehen, mit dem Ziel, den Kraftstoff im Gleitkontakt zu halten und die Verschleißpartikel aufzunehmen. Die Pumpen wurden an einem Pumpenprüfstand unter realistischen Betriebsbedingungen untersucht. Die Versuche wurden mit Isooktan, kommerziellem Ottokraftstoff, Ethanol und Ottokraftstoff/Ethanol-Mischungen durchgeführt. Nach jedem Versuch wurden die untersuchten Komponenten mittels Rasterelektronenmikroskopie (REM) analysiert, um den Verschleiß abschätzen und Reibungsmechanismen identifizieren zu können. Um die Leistung der Versuchsträger zu beurteilen, wurden auch vergleichende Versuche mit zwei Serienpumpen gleicher Bauform durchgeführt.

Die Ergebnisse mit der Einkolbenpumpe und Isooktan als Fördermedium zeigen, dass feingeschliffene Oberflächen im Nocken/Gleitschuh-System den besten Kompromiss aus schnellen Einlauf und niedrigen Reibungszahlen darstellen. Selbstpaarungen aus Siliziumkarbid oder Sialon mit einem α/β-Phasenverhältnis von 60/40 weisen ähnliche Reibungszahlen auf (zwischen 0,02 und 0,05). Im Gegensatz dazu führt die Kombination von Siliziumkarbid mit vergütetem Stahl AISI 52100 über die Laufzeit aufgrund Adhäsionsmechanismen zwischen dem Nocken und dem Gleitschuh zu steigenden Reibungskräften. Adhäsion zwischen Stahl und Siliziumkarbid kann durch eine Textur auf dem Nocken vermieden werden. Jedoch sind die Reibungszahlen mit Isooktan bei niedrigen Drehzahlen deutlich höher als mit selbstgepaartem Siliziumkarbid. Die Textur führt auch zu niedrigeren Reibungsverlusten bei Siliziumkarbid in Selbstpaarung. Wird anstatt Isooktan Eurosuper (ROZ 95) als Fördermedium verwendet, sinken die Reibungszahlen deutlich aufgrund der höheren Viskosität von Ottokraftstoff. Die Reibungszahlen sinken noch weiter mit steigendem Ethanol-Anteil im geförderten Medium. Tribochemische Reaktionen zwischen Ethanol und den untersuchten Werkstoffen könnten zu einer verbesserten Schmierung beitragen, da bereits ein niedriger Ethanol-Anteil im Kraftstoff eine deutliche Verbesserung zur Folge hat. Die niedrigsten Reibungszahlen, die im Nocken/Gleitschuh mit den besten Werkstoffpaarungen erreichbar sind, führen zu einer geringen Beanspruchung im Kolben/Zylinder-System. Die Reibungsverluste mit Siliziumkarbid in Selbstpaarung oder Siliziumkarbid kombiniert mit vergütetem Stahl im Kolben/Zylinder-System weisen Werte auf, die ca. 1,5 % der effektiv geleisteten Pumpenarbeit betragen. Die Analyse der Komponenten, die im Nocken/Gleitschuh-System untersucht wurden, zeigt, dass die Oberflächen geglättet wurden. Jedoch wurde kein maßgeblicher Materialverlust beobachtet. An Kolben und Zylinder wurde ebenso kein erheblicher Verschleiß festgestellt.

Die Untersuchungen mit der Dreikolbenpumpe und Ottokraftstoff als Fördermedium bestätigen die Ergebnisse mit der Einkolbenpumpe. Da keine wesentlichen Unterschiede zwischen den beiden Materialpaarungen, die mit der Einkolbenpumpe im Kolben/Zylinder System eingesetzt wurden, beobachtet wurden, wurde in der Dreikolbenpumpe ausschließlich Siliziumkarbid in Kombination mit vergütetem Stahl AISI 52100 untersucht. Mit selbstgepaartem Siliziumkarbid (ohne oder mit Textur) oder Sialon (mit einem α/β-Phasenverhältnis von 60/40) in den Nocken/Gleitschuh-Systemen konnte ein mechanischer Wirkungsgrad von über 0,95 erreicht werden. Im Vergleich zu untexturierten Teilen erweitert eine Textur auf den Nocken aus Siliziumkarbid den Bereich hohen Wirkungsgrades im Pumpenkennfeld. Texturiertes Siliziumkarbid in Mischpaarung mit AISI 52100 weist ebenso einen sehr guten mechanischen Wirkungsgrad (bis 0,95) ohne Adhäsionsmechanismen in den Gleitsystemen über der Zeit auf. Ein höherer Anteil an α-Phase im Sialon führt zwar zu einer höheren Härte, aber die Reibungsverluste steigen. Darüber hinaus wurden Mikrorisse auf den Oberflächen von Sialon-Nocken mit einem α/β-Phasenverhältnis von 90/10 beobachtet. Diese Risse deuten auf Ermüdungsmechanismen hin. Im Gegensatz zu den Werkstoffpaarungen, die auf Siliziumkarbid basieren, weisen die Sialon-Paarungen Geräuschemissionen und zyklischen Schwankungen des Drehmoments auf. Der Einsatz von einem Sialon-Siliziumkarbid-Verbund erlaubt, solche Phänomenen zu vermeiden, führt aber zu höheren mechanischen Verlusten im Vergleich zu Siliziumkarbid oder Sialon. Mit selbstgepaartem Siliziumnitrid wurden hohe Reibungsverluste und Geräuschemissionen beobachtet. Die Kombination von Siliziumnitrid mit vergütetem Stahl führt zu besseren Ergebnissen, aber der mechanische Wirkungsgrad bleibt unterhalb 0,85 im untersuchten Kennfeldbereich. Im Gegensatz dazu weist Siliziumnitrid mit Kohlenstoffnanoröhrchen in Kombination mit vergütetem Stahl sehr niedrige Reibungsverluste, niedrige Geräuschemissionen und geringe zyklischen Schwankungen auf. Die deutliche Verbesserung, die mit Kohlenstoffnanoröhrchen in der Keramik erreicht wird, kann durch die elastische Verformung der Kohlenstoffnanoröhrchen erklärt werden, wodurch die lokalen Belastungen auf der Nockenoberfläche besser verteilt werden. Darüber hinaus zeigt die Analyse mittels Rasterelektronenmikroskopie, dass tribochemische Reaktionen zu einer Verbesserung der Schmierung in Nocken/Gleitschuh-Systemen beitragen können.

Die Versuche mit Serienpumpen weisen sinkende mechanische und volumetrische Wirkungsgrade über der Laufzeit auf. Die Verschlechterung der Ergebnisse ist auf hohe Reibungsverluste zurückzuführen. Die Bewegung des Kolbens wird durch hohe Reibkräfte in den Gleitsystemen eingeschränkt, was den Ansaugprozess der Pumpe beeinflusst. Auf den Oberflächen der Gleitkomponenten sind makroskopische Kratzer und abgelöste Beschichtungen nach nur ein paar Stunden deutlich sichtbar. Keramische Komponenten in den Gleitsystemen von mediengeschmierten Hochdruckpumpen für Ottokraftstoffe erlauben es, diesen hohen Verscheiß zu vermeiden.

Die Steigerung des Förderdrucks von Hochdruckpumpen für Ottokraftstoffe führt zu weiteren Herausforderungen, insbesondere um die volumetrischen Verluste zu verringern. Diese Verluste, die auf interne Leckage und Rückexpansion zurückzuführen sind, beeinflussen den Gesamtwirkungsgrad der Hochdruckpumpe und müssen durch ein größeres Hubvolumen kompensiert werden. Da die Steigerung des Förderdrucks und die Senkung des volumetrischen Wirkungsgrads zu einer Erhöhung der Pumpenantriebsleistung führen, könnten Einspritzdrücke von deutlich über 200 bar negative Folgen auf den Kraftstoffverbrauch des Verbrennungsmotors haben. Beispielhafte Berechnungen haben jedoch gezeigt, dass der gute Gesamtwirkungsgrad der untersuchten Dreikolbenpumpe mit keramischen Komponenten in den Gleitsystemen nicht zu einem Mehrverbrauch führen sollte. Allerdings wurden diese Berechnungen nur für zwei Betriebspunkte durchgeführt. Weitere Untersuchungen mit dem ganzen Antriebssystem (Verbrennungsmotor und Hochdruckpumpe mechanisch angebunden) sind notwendig, um den realen Einfluss eines erhöhten Einspritzdrucks auf dem Kraftstoffverbrauch im ganzen Motorkennfeld beurteilen zu können. Solche Versuche würden Entwicklern helfen, einen Kompromiss zu finden zwischen der Senkung der Schadstoffemissionen, dem Mehrverbrauch durch die höhere Pumpenantriebsleistung und dem zuverlässigen Betrieb der Hochdruckpumpe.

Author's preface

I started working on the project presented in this document in 2009 as a research associate at the Institut für Kolbenmaschinen (IFKM, part of the Karlsruhe Institute of Technology). The study was funded within the Collaborative Research Centre SFB 483 "High performance sliding and friction systems based on advanced ceramics" by the Deutsche Forschungsgemeinschaft, Federal Ministry of Education and Research, Germany. It is almost impossible to establish an exhaustive list of all the people who contributed to the achievements described in this document. However, I would like to mention some of them in the following lines.

Prof. Dr.-Ing. Ulrich Spicher gave me the opportunity to do my PhD under his mentoring. Thanks to him, I could work with a lot of freedom and autonomy on this very interesting project. I am also grateful to Prof. Dr. rer. nat. Michael J. Hoffmann, who accepted to be my co-mentor and was always available for answering my questions. During my PhD, I could count on the technical and human support of my two successive superiors Dr.-Ing. Sören Bernhardt and Dr.-Ing. Heiko Kubach. Even if he left the IFKM before I arrived, my predecessor Dr.-Ing. Jan Patrick Häntsche still gave me some usefull hints. I especially appreciated the efficient teamwork with Dr.-Ing. Marco Riva and Dr.-Ing. Markus Wöppermann. The many meetings (official or not) I had with them within the SFB 483 were essential to the achievements described in the following pages. I would like to thank Dr.-Ing. Johannes Schneider, Claudius Wörner and Mario Mann who helped to analyse the surfaces of the worn components by providing their expertise. I would also like to express my gratitude to Dr. Manuel Belmonte who developed and manufactured the silicon nitride samples. The investigations were performed with prototype pumps which would not exist without the outstanding work carried out by Ernst Hummel, Axel Vollmer, Ruben Klumpp and Georgij Akkalejew. They converted an innumerable amount of production drawings into real components with high precision. I am grateful to Dominic Creek who cared about the surface finish of the investigated ceramic samples. The assembly of all the components lead to the prototype pumps, but this was not enough to perform investigations. Therefore I would like to thank the mechanics, electricians and electronic technicians from the IFKM who helped to build the test bench. My special thanks goes to Helge Rosenthal and Thorsten Ledig who provided an outstanding support. Many thanks to Florian Schumann and Dr.-Ing. Steve Busch who proofread this document, but also to all my colleagues from the IFKM. I appreciated to work, have some beer and eat some strange french food with them.

Karlsruhe, Juli 2013   Christophe Pfister

Contents

Vorwort des Herausgebers   i

Zusammenfassung   iii

Author's preface   vi

1   Introduction   1

2   Gasoline direct injection engines   5
   2.1   History   5
   2.2   Principle   6
      2.2.1   Operation strategies   6
      2.2.2   Components of the fuel supply system   9
   2.3   Mixture formation mechanisms   10
      2.3.1   Jet break-up and droplet formation   10
      2.3.2   Droplet size and droplet lifetime   12
   2.4   Potential of high-pressure injection   13
      2.4.1   Influence of the injection pressure on the fuel droplet size   .   13
      2.4.2   Increase of the engine efficiency   13
      2.4.3   Effect of high injection pressure on pollutant emissions . . .   14
      2.4.4   Issues related to high injection pressure   14

3   Liquid pumps   17
   3.1   Pump designs   17
   3.2   Reciprocating displacement pumps   19
      3.2.1   Kinematics   19
      3.2.2   Piston pump process   21
      3.2.3   Power and efficiency   22
      3.2.4   Multi piston pumps and pressure impulses   25
   3.3   Specific requirements for gasoline high-pressure fuel pumps   27

4   Tribology   31
   4.1   Definition and challenges   31
   4.2   Tribotechnical system   31
      4.2.1   Engineering surfaces   32
      4.2.2   Contact process   34
      4.2.3   Kinematics   35

ix

4.3   Friction 36
     4.3.1   Friction mechanism 37
     4.3.2   Friction measurands 38
4.4   Wear 38
4.5   Lubrication 41

5   Advanced ceramics   **45**
5.1   Basics 45
     5.1.1   Definition and general properties 45
     5.1.2   Types of advanced ceramics and their application 46
     5.1.3   Manufacturing process 47
     5.1.4   Specific mechanical properties of ceramics 47
     5.1.5   Design of ceramic components 49

6   Investigations   **51**
6.1   Investigated high-pressure fuel pumps 51
     6.1.1   Single-piston prototype pump 51
     6.1.2   3-piston prototype pump 54
     6.1.3   Commercially available high-pressure pumps 57
6.2   Material tested in the sliding systems 58
     6.2.1   Ceramic materials 58
     6.2.2   Material combinations 61
     6.2.3   Material properties 62
     6.2.4   Surface finish 63
6.3   Fuels 64
6.4   Test bench 65
6.5   Test procedure 66
     6.5.1   Single-piston prototype pump 66
     6.5.2   3-piston prototype pump 67
     6.5.3   Commercially available pumps 68

7   Results   **69**
7.1   Performance of the reciprocating sliding systems in the single-piston
     prototype pump 69
     7.1.1   Comparison of the theoretical and measured parameters . . 69
     7.1.2   Friction in the investigated sliding systems 73
     7.1.3   Surface analysis 82
7.2   Application in a 3-piston pump 85
     7.2.1   Comparison of theoretical and measured parameters 85
     7.2.2   Mechanical efficiency with various material pairs in the cam/pusher
           system 86
     7.2.3   Cyclic variations and noise emission 91
     7.2.4   Evolution of the efficiencies over time 95
     7.2.5   Surface analysis 96

7.3 Performance of commercially available pumps 102
    7.3.1 Mechanical efficiency and cyclic variations 102
    7.3.2 Evolution of the Bosch CP1 efficiency over time 104
    7.3.3 Wear analysis 104

8 Discussion   **107**
8.1 Appropriate material pairs for the application in gasoline-lubricated sliding systems 107
    8.1.1 Friction losses 107
    8.1.2 Wear acceptance 108
    8.1.3 Noise emission and cyclic variations 110
    8.1.4 Design complexity and costs 111
    8.1.5 Qualitative evaluation of the investigated material pairs in the cam/pusher system 111
8.2 Optimization of the pump specifications for an improved mechanical efficiency 113
8.3 Volumetric losses 113
8.4 Impact on the engine performance 116

9 Summary   **123**

1 Introduction

Environmental and geopolitical framework

Since their invention, internal combustion engines have been employed in various fields, such as personal mobility, merchandise transportation, hand-held equipment, or electricity production. At its origin, the research and development at combustion engines was focused on increasing the performance and the reliability. However, internal combustion engines are now confronted with new challenges for several reasons, which are listed below.

Predictions regarding fossil energy reserves, even if they are regularly revised, indicate a shortage for 2065 [14]. Moreover, the economic growth and the increasing energy demand of developing countries amplify the problem of fuel supply in the near future. China, for example, has tripled its gross domestic product within the last ten years and will probably overtake the USA (today's largest consumer) regarding energy consumption [58]. The dependency on unstable regions, which own a significant part of the worldwide known resources, present further difficulties as geopolitical conflicts lead occasionally to oil and gas price increases. Between 1975 and 2011, the barrel price (Europe Brent) has increased from approximately $10 /bbl to more than $100 /bbl [1]. Consequently, particular attention is devoted to fossil energy consumption and resources in order to guarantee a sufficient and continuous energy supply in the future, as well as to reduce the dependency on unstable regions of the world. Another major aspect of fossil energy sources is the carbon dioxide (CO_2), which is generated by the combustion of hydrocarbons. Various reports show that the CO_2 emissions due to human activity have an influence on the global climate. According to the report from the Intergovernmental Panel on Climate Change (IPCC), an augmentation of the CO_2 concentration in the atmosphere contributes to global warming and will probably cause environmental disasters such as an increase of the sea level, droughts, heat waves, heavy precipitations, and increased tropical cyclone activity. Furthermore, CO_2 can also directly impact living species since it acidifies the water. All these consequences may impact the ecosystems and threaten the health of millions of people [62].

The geopolitical and environmental framework leads to a strong motivation to reduce the fossil fuel consumption of internal combustion engines. Depending on the application, various alternatives are considered. For personal mobility, electric powertrains are frequently mentioned by the press and the politics. Electric vehicles and batteries are subjected to intensive research. Their development is supported by the European Investment Bank, which recently accorded funds to the Bolloré Group (130,000,000 € in 2011), Nissan Group (220,000,000 € in 2011) and Renault Group (180,000,000 € in 2012) [5, 6, 7]. However, as electricity is only an energy carrier and not a source of energy, this strategy only shifts the problem of CO_2 emissions and energy source from the vehicle to the power plant. Some studies about direct and indirect CO_2 emissions caused by traffic have shown that electric powertrains would generate more emissions than internal combustion engines with the current energy mix in Germany [29, 81]. In addition, electric vehicles still require intensive development in order to provide a competitive alternative to conventional powertrains regarding performances and costs [12]. The cost and the driving range of electric vehicles represent a major disadvantage in comparison to vehicles with combustion engines. As a result, this technology will only concern minor applications in the upcoming years. Only hybrid vehicles (with a combustion engine as main powertrain) seem to be able to gain significant market share in the near future [27].

Concerning the electricity production (a key factor for electric powertrains), the alternative technologies to fossil fuels such as coal, oil, or gas are also subjected to criticism. Nuclear power does not directly produce CO_2 in its energy conversion process but leads to a problem of radioactive waste storage and represents a severe danger for the population in the case of a major accident. Renewable energy sources such as wind or solar energy cause difficulties due to their lack of regularity in power generation.In addition, their low energetic density requires large surfaces to produce the same amount of energy as a nuclear or fossil fuel plant. This issue also prevents the use of renewable energy for commercial transportation (by land and sea), and combustion engines will likely remain the only viable power source for this application for the next decades (apart from mobile nuclear reactors such as those used in some military or maritime ships).

An analysis of the current and future potential performances of internal combustion engines has shown that the efficiency of vehicles (including the energy demand for comfort and safety functions) with combustion engines could reach approximately 65 % whereas electric vehicles may only reach circa 36 % (by considering the system "well-to-wheel") [81]. Consequently, fossil energy sources and bio fuels (which are expected to replace fossil fuels) will probably remain a major energy source for many applications in the next decades.

Potential of gasoline combustion engines

Gasoline and Diesel engines are the two main powertrains deployed in passenger vehicles in Germany with a ratio of about 71 % and 28 % respectively. Gas, hybrid and electric powertrains represent approximately 1 % of the passenger cars [45]. It appears unlikely that one of the two major types of internal combustion engines will replace the other completely as each combustion process offers different advantages and disadvantages regarding fuel consumption, costs, and driving experience. Furthermore, even if the production of Diesel and gasoline fuels are parallel processes from the same base crude oil, they are produced from different sub-products of the oil distillation [88]. Consequently, producing and using both fuels leads to a better exploitation of the crude oil potential.

The research in Diesel engines focuses mainly on the reduction of pollutant emissions such as nitrogen oxides or soot, whereas the research in gasoline engines is essentially focused on the reduction of the fuel consumption. One of the most promising technologies to reduce the fuel consumption of gasoline engines is spray-guided direct injection (described in chapter 2). This technology enables fuel economy improvements of between 5 % and more than 50 %, depending on the engine load and speed in comparison to conventional engines (with port fuel injection) [82]. However, injecting the fuel directly into the combustion chamber causes new difficulties. One of these is an increase of the particulate emissions in comparison to port fuel injection engines [65, 96]. Because of the carcinogenic properties of soot, the emission for passenger cars is limited to 5 mg/km by the european emission standard Euro 5. The future standard Euro 6 to be introduced in 2014 will limit the emission to 4.5 mg/km and the particulate number to $6 \cdot 10^{11}$ #/km [64]. In order to mitigate the increase of particulates emissions that accompany direct injection technology, various strategies can be considered. Using a particulate filter like those used for Diesel engines is one of the possibilities, but implies a more complex and expensive exhaust gas after-treatment. The particulate filter also induces a higher exhaust backpressure and has a negative impact on the engine efficiency [65]. Another way to prevent higher soot emissions is to increase the fuel injection pressure significantly. By improving the mixture formation, the soot emissions are considerably decreased and further improvements in fuel economy can be achieved (more details are given in section 2.4) [16, 17, 74]. However, a higher injection pressure also leads to new difficulties, especially concerning the fuel injection system, which is subjected to significantly higher mechanical and tribological stresses.

One of the components which is most subjected to a stress increase due to higher injection pressures is the high-pressure gasoline pump. Due to technical and economical reasons, the sliding systems of fuel pumps are generally lubricated by the fuel itself (at least the piston/cylinder system). This concept enables a delivery pressure of approximately 250 MPa with Diesel fuel. Such pressure levels cannot be achieved with conventional materials and gasoline lubrication, as gasoline's viscosity is much lower than that of Diesel fuel. The components within the high-pressure pump are subjected to important tribological stresses, thus leading to high friction losses and severe wear. As a result, the durability of the pump is considerably limited. Gasoline high-pressure pumps available on the market can deliver fuel at up to 20 MPa [22, 31] whereas investigations performed at the Institut für Kolbenmaschinen (part of the Karlsruhe Institute of Technology) have shown that a much higher pressure level is required in order to fully exploit the potential of spray-guided direct injection engines [16, 17, 74]. Consequently, the use of advanced materials with enhanced properties (good wear and corrosion resistance, low friction coefficients) appears to be necessary to increase the delivery pressure of gasoline high-pressure pumps.

Research framework and objectives

Advanced ceramics provide properties that may fit an application in the tribotechnical systems of gasoline high-pressure pumps. Ceramics offer a higher hardness than metallic materials and, depending on the ceramic type, also provide good chemical stability with gasoline or ethanol. As part of the Collaborative Research Center SFB483 "High Performance Sliding and Friction Systems Based on Advanced Ceramics" at the Institut für Kolbenmaschinen (KIT), investigations have been performed in order to develop a prototype gasoline pump based on advanced ceramics. These investigations enabled demonstration of the potential of ceramics for this application. The work provided by Häntsche in [37] resulted in the choice of various ceramics showing low friction coefficients and low wear rates when subjected to the tribological stresses that occur in a high-pressure pump. The single piston pump used for these experiments has been in continual development in order to reach a delivery pressure of 50 MPa. Parallel to the work carried out at the Institut für Kolbenmaschinen, research has been performed at the Institut für Keramik im Maschinenbau (IAM-KM) and at the Institut für Werkstoffkunde (IAM-AWP) at the KIT with the objective to optimize the mechanical properties of the materials and the integration into the system.

The investigations described in this document aim at determining which material combinations, operating conditions and geometry enable increases in the injection pressure in comparison relative to modern, commercially available injection systems. According to the most recent research on high-pressure injection in gasoline engines, the target delivery pressure was set to 80 MPa [15, 16, 17, 44, 73, 74, 75]. This pressure level should be reached with a good mechanical efficiency, stable performance and acceptable wear. This performance must be guaranteed with gasoline but also with ethanol as a bio-fuel. In order to achieve these objectives, experiments have been performed on a single piston pump and on a more compact triplex pump. Each of these pumps offers different characteristics and enables measurement of different parameters. Further details concerning the experimental pumps are given in chapter 6.

Contents of this document
The work carried out in this project deals with various scientific aspects which will be basically described in the following chapters. In order to better understand the role of the high-pressure pump, chapter 2 describes the gasoline direct injection technology and shows the potential of increased injection pressures. Chapter 3 discusses the required basics concerning liquid pumps. It focuses essentially on the pump type that was used for the investigations presented in this document and emphasizes the particularities of fuel pumps. Chapter 4 is dedicated to tribology, one of the major aspects of the present experiments. A description of the systems tested, the experimental methods and analysis methods is provided by chapter 6. The results of the investigations are given in chapter 7, and chapter 8 provides a discussion about the results.

2 Gasoline direct injection engines

2.1 History

Even if the share of gasoline direct injection (GDI) engines on the market is increasing since about fifteen years, the idea of injecting the fuel directly into the combustion chamber is much older. Nikolaus Otto already patented this concept in 1877. However, the first application in an internal combustion engine occurred almost 40 years later. In 1916, the Junkers company used gasoline direct injection engines for the first time in speedboats. Afterward, this technology has also been used in aircrafts by Junkers and BMW in 1937. The use of gasoline direct injection led to several advantages in comparison to the conventional technology used at that time (carburetor). It enabled for example a better scavenging, a better combustion chamber cooling and an increase of the compression ratio. As a result, this technology allowed to increase the engine power significantly. The first famous application of gasoline direct injection in series is the introduction of the 300 SL from Daimler-Benz in 1954. The use of direct injection enabled an increase of the engine power by about 10 %. The injection pressure used in the 300 SL was approximately 4.5 MPa.

The development of this technology has been dramatically slowed down due to the relatively low fuel costs and to the development of the port fuel injection (PFI) technology. Some companies such as Texaco, MAN, Ford and Mitsubishi presented several engine concepts with stratified direct injection from 1949 to 1995. The objective was to reduce the fuel consumption and the pollutant emission of reciprocating engines. However, none of these concepts has been produced in series as the tuning of the injection, the mixture formation and the combustion only allowed to operate the engine in stationary points.

It was not until the 1990s that gasoline direct injection was produced in series again in passenger cars. Mitsubishi started the production of the Carisma GDI in 1997. In contrast to the Daimler-Benz 300 SL, the Carisma could be operated in stratified mode (see subsection 2.2.1) in order to reduce its fuel consumption, in particular at part load. The injection pressure of the Carisma engine has barely increased in comparison to the 300 SL and was set to 5 MPa [4]. Afterward, other automotive companies introduced new engines with stratified direct injection such as BMW, Mercedes, Peugeot, Toyota or Volkswagen with an injection pressure of up to 20 MPa [26, 35, 76, 90, 94, 97]. Despite its lower fuel economy improvement, most of the companies then switched to homogeneous direct injection concepts because of the high development costs and exhaust gas after-treatment issues induced by stratified operation [9, 13, 23, 24, 31, 36, 43, 46, 48, 49, 84, 91].

2.2 Principle

2.2.1 Operation strategies

The use of the direct injection technology in gasoline engines offers several advantages depending on the operating strategy. The figure 2.1 shows the most common operating modes in use as a function of the engine operating point (engine speed and load). At part load and low engine speed, the engine may be operated in stratified mode with lean mixture (air-fuel ratio $\lambda \gg 1$). At higher engine speed and higher load, the engine may be operated in lean homogeneous mode ($\lambda \leq 1.5$), or stoichiometrical homogeneous mode ($\lambda = 1$) at the highest loads and speeds. In some cases, the mixture is enriched ($\lambda < 1$), for example in order to protect the engine components from overheating by reducing the combustion and exhaust gas temperature. Depending on the operating point considered, the fuel economy improvement can reach more than 50 % in comparison to an equivalent PFI engine [9, 82].

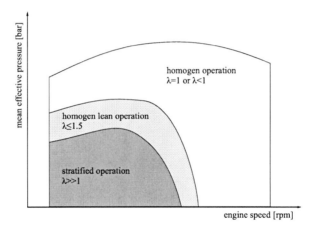

Figure 2.1: Typical operating map of a gasoline direct injection engine [9]

Homogeneous operation
In homogeneous mode, the fuel is injected during the intake stroke and is mixed with the air flowing into the combustion chamber. The time available for mixture formation and the air motion during aspiration support the formation of an almost homogeneous mixture before ignition.

At full load (stoichiometrical mixture), the benefit of direct injection is relatively low when compared to an equivalent PFI engine (see figure 2.2). Injecting the fuel directly into the combustion chamber leads to a mixture cooling, thus increasing the volumetric efficiency and reducing the knock tendency. As a result, the compression ratio can be increased by about $\Delta\varepsilon \approx 1.5..2$, and a better thermal efficiency can be achieved. Due to the improved volumetric and thermal efficiency, the benefit of direct injection at full load is approximately 2 % to 5 % [9, 24, 82].

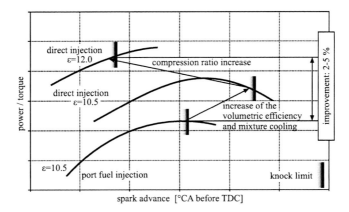

Figure 2.2: Improvement through the use of direct injection at engine full load [24]

The homogeneous mode can be used over the entire operating range of a GDI engine, including part load. However, as the ignition limit of the air-fuel mixture is reached at about $\lambda \approx 1.5$, the engine load has to be set via throttling in order to keep the air-fuel ratio below the ignition limit. As a result, the potential of homogeneous operation is lower than stratified operation regarding fuel economy because of the throttling losses during the intake. In turn, the homogeneous operating strategy provides more time for mixture formation, which means a simpler application in comparison to stratified operation. Moreover, if the engine is operated with a stoichiometrical mixture, it is possible to use a conventional 3-way-catalyst to reduce the pollutant emissions [9, 53].

Stratified operation
In order to run the engine without throttling and thus to reduce the pumping losses, the engine has to be operated with a lean mixture (air-fuel ratio $\lambda \gg 1$). But as mentioned before, the ignition limit of the air-fuel mixture is about $\lambda \approx 1.5$. The stratification strategy enables to extend the ignition limit by providing an inhomogeneous charge with a local rich and ignitable mixture near the spark plug but a globally lean mixture in the entire combustion chamber (up to $\lambda > 8$). Through the significant reduction of the pumping losses, this operating mode offers a higher potential regarding fuel economy at part load in comparison to the homogeneous mode [9, 95].

The stratification of the air-fuel mixture is realized by injecting the fuel during the compression stroke, before ignition. Two successive generations of direct injection engines have been developed to achieve this objective. Figure 2.3 shows the different concepts used in the first generation of stratified GDI engines. In wall-guided injection concepts, the injected fuel spray is redirected via the piston surface with a specific geometry towards the spark plug. The air motion inside the combustion chamber (swirl, tumble) supports the mixture formation. Engines with air-guided fuel injection use the air motion inside the cylinder to transport the ignitable mixture to the spark plug. Actually, these two stratification concepts are closely linked as the geometry of the piston influences the air motion and the air motion supports the mixture formation. The injectors used in the first generation of GDI engines were mainly swirl injectors, which provide very small droplets at an injection pressure of 5 MPa to 12 MPa. However, the spray generated by these injectors is very sensitive to air motion and pressure variations in the cylinder. This means that satisfactory engine tuning can only be achieved in a narrow range of the engine operating map. The spray stability is not satisfactory at high engine speed (typically above 3,000 rpm) and an ignitable mixture near the spark plug at ignition timing is not guaranteed. Moreover the engine load which can be achieved with fuel stratification is limited because of the soot production in the regions of incomplete mixture preparation. Another issue is the piston-wetting, especially in the case of wall-guided fuel injection. As a part of the injected fuel gets in contact with the piston surface and do not vaporize, an incomplete combustion can occur and cause high unburned hydrocarbons and soot emission in the exhaust. Due to these difficulties, the potential of GDI engines is neutralized and therefore a second generation of stratified GDI engines has been developed [24, 53, 56, 78, 82].

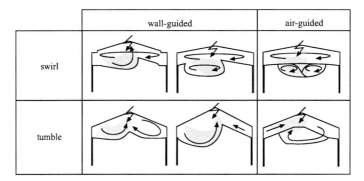

Figure 2.3: First generation of direct injection engines with stratified mode: wall-guided and air-guided concepts [24]

The second generation of GDI engines uses a spray-guided injection by placing the injector near the spark plug as shown in figure 2.4. Due to the narrow spacing of the injector and the spark plug, the time available for fuel vaporization is significantly reduced in comparison to the first generation of stratified GDI engines. In order to guarantee a good mixture preparation, the injection pressure of spray-guided engines is increased to up to 20 MPa. The injectors used for this concept are outwards-opening piezo injectors with a higher spray stability than swirl injectors. When compared to the first generation of GDI engines, spray-guided GDI engines have an extended stratified operation range, thus enabling to better exploit the benefit of direct injection. However this concept leads to several difficulties. Like wall-guided or air-guided stratification, an ignitable mixture must be located near the spark plug at ignition timing. As the time available for mixture preparation is shorter than in the case of wall-guided or air-guided concepts, the stratification is intensified and the position of the components such as the injector and the spark plug must be set very precisely. The short time between injection and ignition can also lead to a insufficient fuel vaporization, thus increasing the soot and unburned hydrocarbon emissions. These emissions are the main limits of the stratified mode operating range. Moreover, the spark plug is subjected to an high thermal stress as it is heated by the combustion and cooled down by the fuel during the injection very rapidly [24, 56].

Figure 2.4: Second generation of direct injection engines with stratified mode: spray-guided concept [53]

The mixture stratification enables to reduce the pumping losses significantly by operating the engine with wide open throttle. However, it also leads to new challenges regarding pollutant emissions as these reduce the operating range of stratification. An increase of soot and hydrocarbon emissions due to the short time available for mixture formation has already been mentioned. But other difficulties occur because of the lean mixture operation. For example, the excess of oxygen prevents the conversion of nitrous oxides by a conventional 3-way-catalyst. Moreover the exhaust gas temperature is lower and may not reach the catalyst light-off temperature [9, 95].

2.2.2 Components of the fuel supply system

The most evident difference between the design of PFI and GDI engines is the implementation of the injectors. Another major difference is the fuel supply system. Figure 2.5 shows the main components to be found in common injection systems. The PFI engines have a relatively simple fuel system with an electrical pump feeding gasoline to the injectors at a pressure of approximately 0.4 MPa. Depending on the technology used, the pump is driven permanently at full load (with return pipe to the tank for the excessive fuel) or with flow rate regulation (no return pipe required). For direct injection, a higher injection pressure is necessary to compensate the very short time available for mixture preparation in comparison to PFI engines. This leads to a more complex fuel supply system as shown in figure 2.5 (on the right). An electrical feed pump supplies fuel at approximately 0.4 MPa to the high-pressure pump, which is mechanically driven by the combustion engine. The high-pressure pump increases the pressure to up to 20 MPa. The fuel is delivered to the injectors via the common rail, like in modern Diesel injection systems. The fuel pressure is set by a pressure regulation valve integrated in the rail or directly in the high-pressure pump [9, 31, 68].

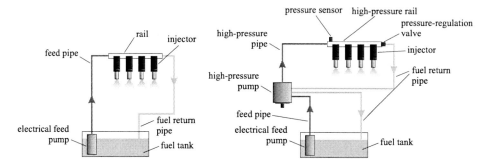

Figure 2.5: Main components of the fuel supply systems: port fuel injection (left) and direct injection (right)

2.3 Mixture formation mechanisms

The gasoline direct injection (in particular the stratified mode) offers a strong potential to reduce the fuel consumption in comparison to PFI engines. However, it leads to a great challenge regarding the mixture preparation. As explained in the subsection 2.2.1, the time available for mixture preparation is very short. If the mixture located near to the spark plug is too lean, it cannot be ignited. If it is too rich, a high soot formation occurs. Moreover, if the spray reaches the walls of the combustion chamber, the fuel may not vaporize before ignition, thus causing unburned hydrocarbon emission. As a result, the injection process has to be tuned very precisely in order to ensure good spray stability and low pollutant emission levels.

2.3.1 Jet break-up and droplet formation

The mixture preparation begins with the injection of the fuel into the combustion chamber through the injector nozzle. Depending on the jet speed, various mechanisms of break-up can be observed. The type of jet break-up directly influences its length as shown in figure 2.6. At very low velocities (A to B on picture 2.6), drip flow occurs. In the Rayleigh regime, the jet length increases with increasing velocity. The surface tension forces dominate and the jet breaks up in droplets, which are bigger than the nozzle hole diameter. At higher jet velocity, the jet length decreases as a result of the aerodynamic forces and the droplet mean diameter is in the range of the nozzle diameter. This regime is called first wind-induced break-up. A further increase of the jet velocity leads to the second wind-induced regime with a turbulent flow in the nozzle. In this regime, the droplet size is smaller than the nozzle diameter. If the intact surface length of the jet tends to zero, the atomization regime is reached and a conical spray is formed. As the diameter of the droplets is much smaller than the nozzle diameter, this regime is relevant for gasoline injection in combustion engines [9, 11]

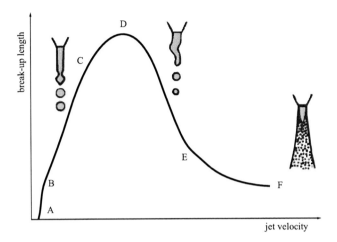

Figure 2.6: Jet surface break-up length and regime as a function of the jet velocity; A→B: drip flow, B→C: Rayleigh break-up, D→E: first wind-induced break-up, E→F: second wind-induced breakup, beyond F: atomization regime [11]

The break-up regime can be characterized by the Weber number *We*, which relates the dynamic pressure to the internal pressure of the liquid. It is defined as a function of the air density ρ, the relative speed between the air and the injected fuel c_{rel}, the diameter of the droplets D and the surface tension of the fuel σ_{fuel}:

$$We = \frac{\rho_{air} \cdot c_{rel}^2 \cdot D}{\sigma_{fuel}} \tag{2.1}$$

The break-up of gasoline in air occurs when the Weber number exceeds its critical value of approximately $We_{critical} \approx 10..12$. The equation 2.1 shows that the jet break-up can be influenced by increasing the air density (increasing the combustion chamber pressure via late injection or supercharging for example) or by increasing the relative speed between the injected fuel and the surrounding air. Larger droplets also lead to a higher Weber number, resulting in a break-up into smaller droplets, until their Weber number remains lower than the critical value $We_{critical}$ [9, 11].

Figure 2.7 gives a schematic view of the spray break-up during the injection of gasoline via a multi-hole injector (similar mechanisms occur in the case of outward-opening injectors). Two regions can be identified. In the region of the primary break-up (at nozzle exit), the jet is mainly composed of liquid and is broken up into droplets. In the secondary break-up region (at a greater distance from the nozzle), the droplets are broken-up into even smaller droplets by aerodynamic forces. The droplets then vaporize via heat exchange with the surrounding air [9, 11].

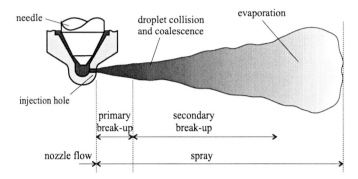

Figure 2.7: Structure of a full-cone high-pressure spray [11]

2.3.2 Droplet size and droplet lifetime

In order to accelerate the mixture formation, a fast vaporization of the injected fuel is required. Figure 2.8 shows the lifetime of a fuel droplet as a function of its diameter and the temperature of the surrounding air. This diagram indicates that the vaporization is faster with a higher air temperature (higher heat supply) but also with smaller droplets (heat exchange support through the increase of the surface to volume ratio). As a result, reducing the droplets diameter by 50 % enables to reduce their lifetime by approximately 75 % [3].

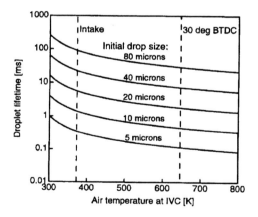

Figure 2.8: Effect of droplet size and air temperature on droplet lifetime [3]

As the injected fuel droplets do not all have the same size, it is common to mention the Sauter mean diameter (SMD), which best characterizes the surface to volume ratio.

The Sauter mean diameter D_{32} is defined as a function of the quantity N_i of droplets of a given diameter D_i via the following equation [50]:

$$D_{32} = \frac{\sum^i N_i \cdot D_i^3}{\sum^i N_i \cdot D_i^2} \tag{2.2}$$

2.4 Potential of high-pressure injection

The potential of an increased injection pressure in comparison to the current standard of 20 MPa has been analyzed in various investigations. By increasing the pressure to 100 MPa, many advantages could be demonstrated. The main benefits are an accelerated injection combined to an improved mixture preparation (essential for stratified operation), a faster combustion and decreased hydrocarbons, nitrous oxides and soot emissions [16].

2.4.1 Influence of the injection pressure on the fuel droplet size

The main explanation of the benefits of high-pressure injection listed above is the increased speed of the fuel jet at the injector nozzle exit. This leads to a higher Weber number and results in an intensification of the jet break-up. Figure 2.9 shows the Sauter mean diameter of droplets injected into a pressure chamber with a back pressure of 0.4 MPa to 1.6 MPa. The injection pressure is increased from 20 MPa to 50 MPa and leads to a decrease of the Sauter mean diameter of 30 % to 40 % depending on the back-pressure in the chamber. As the droplets are smaller, their surface to volume ratio increases and their vaporization is accelerated (as shown in figure 2.8). Furthermore, the higher jet speed leads to a higher fuel flow rate and enables a shorter injection. As a result, the whole injection and mixture preparation process is improved by the high injection pressure [44, 59, 60, 63, 77].

2.4.2 Increase of the engine efficiency

By increasing the injection pressure from 20 MPa to 100 MPa in a GDI engine operated in stratified mode, a reduction of the specific fuel consumption of approximately 4 % can be achieved. This benefit can be explained by several effects. The air-fuel mixture realized in stratified mode with high-pressure injection is more compact and leads to a faster combustion. Moreover, the acceleration of the injection and mixture formation processes provides more flexibility for the tuning of the injection and ignition parameters. As a result, the 50 % mass fraction burned timing can be set at optimal crank angle (around 8 °CA after top dead center) and the engine efficiency is increased. Furthermore, the compact mixture leads to lower heat losses during the combustion as it is insulated by the surrounding air. However, the higher driving power required for the high-pressure pump is not taken into account in the benefits listed above [9, 15, 16, 17, 44, 56, 59, 63].

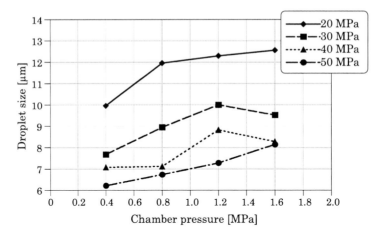

Figure 2.9: Influence of the injection pressure and chamber pressure on the droplet size (SMD) by using a multi-hole injector [44]

2.4.3 Effect of high injection pressure on pollutant emissions

A major advantage provided by an increased injection pressure is the reduction of the pollutant emissions. The improved mixture preparation leads to a significant reduction in soot emission. Investigations have shown that increasing the fuel injection pressure from 20 MPa to 100 MPa in a GDI engine operated in stratified mode enables to reduce the soot emission by approximately 90 %. As a result, the engine can be operated with stratification in a wider load range without exceeding the soot emission limit. Moreover the nitrous oxides and hydrocarbon emissions can be reduced by 17 % and 35 %, respectively. The lower nitrous oxide emission can be explained by the strong stratification, which leads to a locally rich combustion. The lower hydrocarbon emission can be explained by a shorter spray penetration, which prevents piston-wetting [15, 17, 44].

The benefits of high-injection pressure can also be used for cold start by operating the engine in stratified mode and igniting beyond the top dead center. This enables a fast heating of both the engine and the catalyst which provides low pollutant emissions and good combustion stability. By increasing the injection pressure to 60 MPa, the indicated fuel consumption during the catalyst heating could be reduced by approximately 24 % with a better combustion stability in comparison to an injection pressure of 20 MPa. In addition, the nitrous oxides, unburned hydrocarbons, carbon monoxide and soot emissions could be reduced by 40 %, 81 %, 58 % and 76 %, respectively [75].

2.4.4 Issues related to high injection pressure

The advantages of an increased injection pressure listed above are linked to several difficulties. The driving power required for the high-pressure pump increases with the injection pressure (see chapter 3). As a result, the benefit of high-pressure injection regarding fuel efficiency may be neutralized or even outweighed by the higher engine power required to drive the pump.

Some papers report about the higher spray penetration which results from increased injection pressure [55, 56]. If the spray reaches the walls of the combustion chamber (cylinder liner or piston), a part of the fuel will not take part to the combustion and leads to a lower efficiency as well as a higher unburned hydrocarbon emission. However, other experiments have shown that the spray penetration only increases in the pressure range of current injection systems. At higher pressure, the spray penetration decreases due to the shorter injection duration and the accelerated vaporization [15, 44, 63].

Another major challenge linked to the high-pressure injection is the increase of the tribological stresses in the high-pressure fuel pump. As the lubricity of gasoline is relatively low, the friction losses in the sliding systems increase and the pump lifetime is reduced. This issue has to be solved in order to enable the application of higher injection pressures in GDI engines.

Symbols

Symbol	Unit	Description
α	[°CA]	crank angle
ε	[-]	compression ratio
λ	[-]	air-fuel ratio
ρ_{air}	[kg/m^3]	air density
σ_{fuel}	[N/m]	surface tension of the fluid
c_{rel}	[m/s]	relative speed between injected fuel and air
D	[µm]	droplet diameter
SMD	[µm]	Sauter mean diameter
N	[-]	number of droplets of a given size
We	[-]	Weber number

3 Liquid pumps

Chapter 2 has shown the potential of an increased injection pressure in GDI engines. The high-pressure pump is a key component to achieve this objective. This chapter gives some basic theory about liquid pumps in order to better understand the investigations presented in this document.

3.1 Pump designs

The general function of a pump is basically to increase the potential and kinetic energy (pressure and flow rate) of a fluid [51]. The pumps are used in most of the technical fields such as electricity production, chemistry, pharmaceutics, metallurgy or transport [66]. Pumps provide a very wide range of power output: a relative pressure of 10^{-2} MPa to 10^3 MPa and a flow rate of $1 \frac{ml}{h}$ to $10^5 \frac{m^3}{h}$ can be realized [47, 51]. Depending on the requirements of the application, various pump types can be used as shown in figure 3.1.

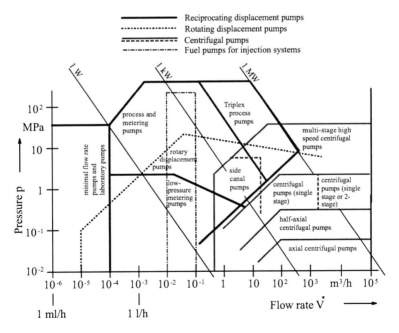

Figure 3.1: Main pump categories depending on their delivery pressure and flow rate [47]

The two main pump families are the centrifugal pumps and the positive displacement pumps. The centrifugal pumps operate as follows: the rotation of a paddle-wheel increases the kinetic energy of the liquid. This energy is transformed into potential energy by exploiting the inertia of the fluid in motion. Centrifugal pumps provide a pulsation-free flow. The pressure and flow rate levels of centrifugal pumps are essentially controlled by the circumferential speed of the paddle-wheel. This leads to an interdependence between the flow rate, the pressure and the rotation speed of the driveshaft as shown in figure 3.2 [51].

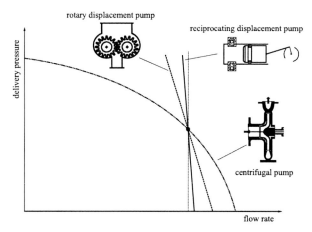

Figure 3.2: Performance curves of centrifugal and displacement pumps at constant rotation speed [47]

Positive displacement pumps aspirate the liquid from the suction inlet pipe and displace it to the outlet pipe via the work chamber. The volume of the work chamber varies as a function of the mobile parts motion. The volume change forces the displacement of the liquid and increases its potential and kinetic energy. In this type of pump, the flow rate is a direct function of the size of the work chamber and the working frequency. The positive displacement pumps family can be split in two groups depending on the kinematics: rotating pumps and reciprocating pumps [87]. In contrast to centrifugal pumps (and to a lesser extent rotary displacement pumps), reciprocating displacement pumps offer the advantage of a very low dependency between flow rate and pressure (see figure 3.2). As a result, the output pressure is almost independent from the rotation speed of the driveshaft. Moreover, this type of pump provides a very precise flow rate and allows to reach very high delivery pressures (up to 10^3 MPa) in comparison to centrifugal pumps [34, 47].

The fuel consumption of an internal combustion engine at full load is almost proportional to its rotation speed. Assuming that a high-pressure fuel pump is driven by the combustion engine and that the injection pressure should be set independently from its rotation speed, reciprocating displacement pumps offer the best characteristics for modern injection systems. Therefore the remainder of this chapter only concerns this pump type.

3.2 Reciprocating displacement pumps

Reciprocating displacement pumps increase the kinetic and potential energy of a fluid via a piston in the work chamber. The piston can be either rigid or soft. Soft pistons such as membranes enable to realize a fully leak-proof work chamber [34]. However membrane pumps are typically used for applications which do not require pressures above 30 MPa and their design is relatively complex [47, 87]. Since the objective of the present work is to provide a pressure of 80 MPa, only rigid piston pumps will be considered in this chapter. Figure 3.3 shows the main components of this pump type, including the piston, the work chamber and the inlet and outlet valves.

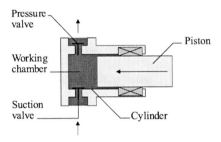

Figure 3.3: Schematic view of a reciprocating displacement pump (piston pump)

3.2.1 Kinematics

In order to describe the evolution of the pressure in the work chamber as a function of the time (or of the driveshaft angle), it is necessary to define the motion of the piston. The piston motion depends on its driving system. This subsection presents two common types of piston driving systems. Figure 3.4 describes the kinematics of a crankshaft/rod system, which is similar to the system used in combustion engines and of an eccentric shaft/pusher system, which is used in several high-pressure fuel pumps [30, 31, 68]. In a convention for this chapter, the piston is at top dead center (TDC) for a driveshaft angle of $\alpha = 0°$ and at bottom dead center (BDC) for $\alpha = 180°$.

In the case of a crankshaft/rod system as shown in figure 3.4, the piston is driven by the rod (length: l), which transforms the rotation of the crankshaft (radius: r) into a reciprocating movement. In the case of an eccentric shaft/pusher system, the piston is directly driven by the eccentric (eccentricity: e) rotating with the driveshaft. In both cases, presented in figure 3.4, x describes the piston travel with $x = 0$ at $\alpha = 0°$ (TDC) as a reference.

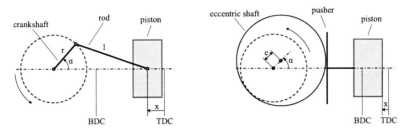

Figure 3.4: Kinematics of reciprocating pumps: crankshaft/rod (left) and eccentric shaft/pusher (right)

Piston travel

According to the conventions used in this chapter, the piston travel in a crankshaft/rod system can be defined as a function of the driveshaft angle as follows [34]:

$$x = r \cdot (1 - \cos \alpha) + l \cdot \left(1 - \sqrt{1 - \left(\frac{r}{l}\right)^2 \cdot \sin^2 \alpha}\right) \tag{3.1}$$

The piston travel in the case of an eccentric shaft/pusher system is given by:

$$x = e \cdot (1 - \cos \alpha) \tag{3.2}$$

Piston speed

The piston speed v is defined as the time derivative of the piston travel x :

$$v = \frac{\mathrm{d}x}{\mathrm{d}t} = \frac{\mathrm{d}x}{\mathrm{d}\alpha} \cdot \frac{\mathrm{d}\alpha}{\mathrm{d}t} \tag{3.3}$$

An approximation of the piston speed with a crankshaft/rod system is given by following equation [34]:

$$v \approx \left[\omega \cdot r \cdot \sin \alpha + \frac{r}{2 \cdot l} \cdot \sin(2 \cdot \alpha)\right] \tag{3.4}$$

with $\omega = \frac{\mathrm{d}\alpha}{\mathrm{d}t}$ designating the angular speed of the driveshaft.

In the case of an eccentric shaft/pusher system, the piston speed is given by the following function:

$$v = \omega \cdot e \cdot \sin \alpha \tag{3.5}$$

3.2.2 Piston pump process

The reciprocating motion of the piston causes an intermittent inlet and outlet flow. The pump operates in two successive strokes:

- the charge cycle (suction valve open, pressure valve closed): the piston moves downwards from the top dead centre to the bottom dead centre and the liquid flows into the work chamber through the suction valve

- the work cycle (suction valve closed, pressure valve open): the piston moves upwards to the top dead centre and the liquid is pumped into the high-pressure pipe via the pressure valve

Figure 3.5 shows the pressure-volume diagram of a reciprocating piston pump. For this example, it is assumed that there is no internal leakage in the pump (this phenomenon will be explained in subsection 3.2.3). The pump process in the case of an incompressible liquid is given by the following path: $1 \rightarrow 2$ (isobaric suction), $2 \rightarrow 3$ (isochore compression), $3 \rightarrow 4$ (isobaric delivery) and $4 \rightarrow 1$ (isochore expansion) [34].

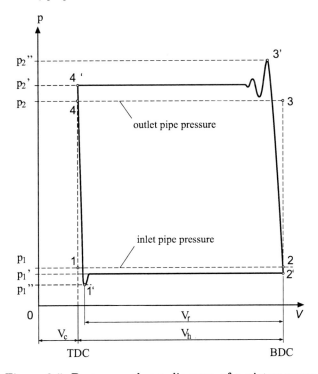

Figure 3.5: Pressure-volume diagram of a piston pump

The variation of the work chamber volume between TDC and BDC is the maximum swept volume (or pump displacement) V_h. It is a function of the maximum piston travel s_p and the piston diameter d:

$$V_h = \frac{s_P \cdot \pi \cdot d^2}{4} \tag{3.6}$$

The cylinder volume V_{cyl} is the sum of the maximum swept volume and the dead volume V_c left when the piston reaches TDC:

$$V_{cyl} = V_h + V_c \tag{3.7}$$

Depending on the delivery pressure and the liquid properties, it can be necessary to take the compressibility of the liquid into account. In this case, the compression and expansion processes are different from the incompressible liquid case. Assuming that the compression and expansion are isothermal, the volume V and the pressure p in the work chamber are linked via the compressibility factor of the fluid β_T as follows [47, 61, 86]:

$$\left| \frac{\triangle V}{V} \right| = \beta_T \cdot \triangle p \tag{3.8}$$

As a result, a delay can be observed between the beginning of the compression (point 2 in figure 3.5) and the reach of the theoretical delivery pressure p_{out}. A similar effect can be observed when the piston starts moving downwards (point 4) and reaches the theoretical suction pressure p_{in}. In contrast to the compression, the expansion of a compressible liquid causes volumetric losses since the actual aspirated volume V_r is lower than the maximum swept volume V_h. Furthermore, the pressure losses in the valves (caused by flow resistance) require a compensation in order to keep the pressure in the inlet and outlet pipes at the desired level. Consequently, the minimum cylinder pressure p_{min} is lower than the pressure in the inlet pipe and the maximum cylinder pressure p_{max} is higher than the outlet pipe pressure. The pressure-volume process in the case of a compressible liquid and with pressure losses in the valves is given in figure 3.5 by the path $1' \rightarrow 2' \rightarrow 3' \rightarrow 4'$ [34].

3.2.3 Power and efficiency

The work W_i and the indicated power P_i provided by the pump to the liquid over one cycle (one single driveshaft revolution) at a rotation speed n is given by following equations :

$$W_i = \oint V \, dp \tag{3.9}$$

$$P_i = W_i \cdot n \tag{3.10}$$

The conversion of mechanical energy into kinetic and potential energy leads to energy losses, which can be classified in different categories. Figure 3.6 gives a schematic view of these losses between the pump input and output.

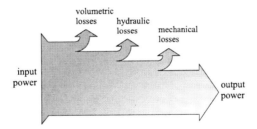

Figure 3.6: Power losses between the input and the output of a piston pump

The input power P_{in} required to drive the pump is a function of the driving torque M_d and the angular speed ω of the driveshaft:

$$P_{in} = M_d \cdot \omega \qquad (3.11)$$

The pump effective power P_{out} is given as a function of the input and output pressure (p_{in} and p_{out}) and the volumetric flow $\dot{V}_{real}$:

$$P_{eff} = (p_{out} - p_{in}) \cdot \dot{V}_{real} \qquad (3.12)$$

Consequently, the pump total efficiency η_{total} can be calculated as the ratio of the output power to the input power as follows [61]:

$$\eta_{total} = \frac{P_{out}}{P_{in}} = \frac{(p_{out} - p_{in}) \cdot \dot{V}_{real}}{M_d \cdot \omega} \qquad (3.13)$$

The energy losses of the pump can be splitted in three categories: the volumetric losses, the hydraulic losses and the mechanical losses. The volumetric losses are characterized by the ratio between the real volumetric flow rate $\dot{V}_{real}$ and the theoretical volumetric flow rate $\dot{V}_{theo}$ of the pump. The theoretical flow rate is a function of the maximum swept volume and the pump rotation speed. The volumetric efficiency η_{vol} is defined as follows [89]:

$$\eta_{vol} = \frac{\dot{V}_{real}}{\dot{V}_{theo}} = \frac{\dot{V}_{real}}{V_h \cdot n} \qquad (3.14)$$

The volumetric efficiency can also be calculated via the real and theoretical mass flow ($\dot{m}_{real}$ and $\dot{m}_{theo}$ respectively) of the pump as shown in equation 3.15. This equation is linked to the equation 3.14 via the density of the liquid ρ:

$$\eta_{vol} = \frac{\dot{m}_{real}}{\dot{m}_{theo}} = \frac{\rho \cdot \dot{V}_{real}}{\rho \cdot \dot{V}_{theo}} \qquad (3.15)$$

The volumetric losses are partially caused by internal leakage. Leakage may occur in the valves but also between the piston and the cylinder. The seal between these two components can be realized for instance by a multi-seal stuffing box or by a very small clearance acting as a throttle (throttle gap seal). The first solution prevents the leakage almost completely. However it causes higher friction between the piston and the cylinder and thus reduces the pump mechanical efficiency [61]. In the case of a throttle gap seal, a leakage flow $\dot{V}_{leak}$ between the piston and the cylinder is tolerated (figure 3.7). The flow rate can be calculated via the Hagen-Poiseuille formula as a function of the nominal piston diameter d, the clearance between the piston and the cylinder c, the pressure difference between the working chamber and the low pressure chamber $(p_{high} - p_{low})$, the cylinder length l_c and the viscosity of the liquid η (see equation 3.16) [61, 89]. In general, the liquid leaking from the work chamber flows back to the suction pipe of the pump.

$$\dot{V}_{leak} = \frac{d \cdot \pi \cdot c^3 \cdot (p_{high} - p_{low})}{12 \cdot l_c \cdot \eta} \tag{3.16}$$

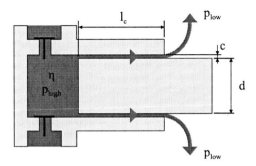

Figure 3.7: Internal leakage between the piston and the cylinder

Apart from the internal leakage, the elasticity of the pump and the compressibility of the liquid also contribute to the volumetric losses. As explained in subsection 3.2.2, the liquid which remains in the dead volume V_c after TDC is expanded before the actual suction begins. This phenomenon is called re-expansion. The dead volume cannot be completely avoided because of the valve design and the required clearance between the piston and the top of the work chamber at TDC. Moreover, if the pump is subjected to high pressure forces, the dead volume increases due to the deformation of the work chamber. An increased dead volume causes a longer expansion during the charge cycle of the pump and reduces the real aspirated volume V_r as described in figure 3.5. The volume lost by the re-expansion of the liquid in the dead volume is given by following equation (assuming an isothermal expansion, see equation 3.8):

$$V_h - V_r = \frac{V_c}{\frac{1}{\beta_T \cdot (p_{max} - p_{min})} - 1} \tag{3.17}$$

The hydraulic efficiency is linked to the friction of the liquid flowing through the pump, especially through the suction and pressure valves (throttling effect). The hydraulic losses can be identified in figure 3.5. The pressure p_2 available at pump outlet is lower than the pressure in the work chamber p_2' during the delivery and the pressure p_1 in the inlet pipe is higher than the work chamber pressure p_1' during the suction. The hydraulic efficiency is given by the pressure difference between the inlet and outlet pipes $(p_{in} - p_{out})$ and the indicated mean effective pressure p_{mi} as shown in equations 3.18 and 3.19:

$$\eta_{hydr} = \frac{p_{out} - p_{in}}{p_{mi}} \tag{3.18}$$

$$p_{mi} = \frac{W_i}{V_h} \tag{3.19}$$

The mechanical losses are caused by the friction between the moving parts of the pump such as the piston or the bearings. The mechanical efficiency is given by following equation [28]:

$$\eta_{mech} = \frac{P_i}{P_{in}} \tag{3.20}$$

The total pump efficiency can be described as the product of the three efficiencies previously mentioned:

$$\eta_{total} = \eta_{vol} \cdot \eta_{hydr} \cdot \eta_{mech} \tag{3.21}$$

Figure 3.8 shows the qualitative evolution of the volumetric, mechanical and total pump efficiencies as a function of the delivery pressure level. The volumetric efficiency decreases linearly with increasing pressure as suggested by equations 3.17 and 3.16. The mechanical efficiency increases with increasing delivery pressure. This can be explained by the low interdependence between friction losses and delivery pressure: the energy lost by friction remains almost constant whereas the energy provided to the liquid increases with increasing pressure [92]. The hydraulic efficiency is not displayed in the diagram as it can be assumed that the pressure losses in the pump remain constant at constant rotation speed and are not affected by the delivery pressure level.

3.2.4 Multi piston pumps and pressure impulses

The previous sections only considered single piston pumps. However, multi piston pumps are used in some applications since they provide several advantages. They enable for example to reach a high flow rate in a small constructed size. Depending on the requirements of the application (especially regarding the pump size), different designs may be used. Figure 3.9 shows some of them. Another important benefit is the significant reduction of the pressure fluctuations in the inlet and outlet pipes [47]. As explained in the subsection 3.2.2, the pump flow rate is intermittent because of the pump kinematics and process. As a result, pulsation occurs in the inlet and outlet pipe and may cause damages or operation disruptions. Consequently, pumps causing pulsations require special care on the pipes and connections designs [31, 80, 87, 89].

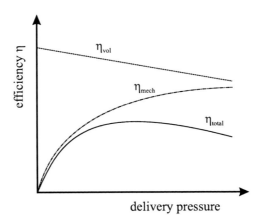

Figure 3.8: Pump efficiencies against delivery pressure (qualitative evolution at constant temperature and speed) [92]

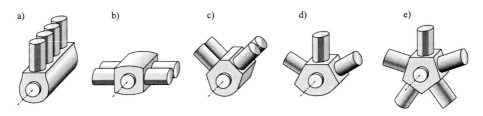

Figure 3.9: Examples of multi piston pump designs: a) in-line pump; b) opposed-cylinder pump; c) V-pump; d) W-pump; e) radial-pump [34]

Figure 3.10 shows the qualitative flow rate of various pump designs over one cycle. It is assumed that the angular phase shift φ between the pistons is set to $\varphi = \frac{360}{z}$ degrees, with z the number of pistons. This means that a triplex pump has an angular phase shift of $\varphi = 120\,°\text{CA}$ (Crank Angle). The instant total pump mass flow $\dot{m}_{total}$ at a time t is given by the sum of the single instant mass flows of each work chamber i: $\dot{m}_{total}(t) = \sum \dot{m}_i(t)$.

In the case of single or dual piston pumps, the relative flow rate varies between $0\,\%$ and $100\,\%$, which means $100\,\%$ pulsation. Triplex pumps cause only $3\,\%$ to $7\,\%$ pulsation (flow rate between $93\,\%$ and $100\,\%$) depending on the piston kinematics and assuming a volumetric efficiency of $\eta_{vol} = 1$. The pulsations can be reduced even more with five pistons ($2\,\%$ to $3\,\%$ pulsation). However, if the volumetric efficiency decreases, the pulsations increase. For example, if the volumetric efficiency of a triplex pump is lower than 0.7, the pulsations in the pressure pipe reach a level of $100\,\%$ [28, 47, 89].

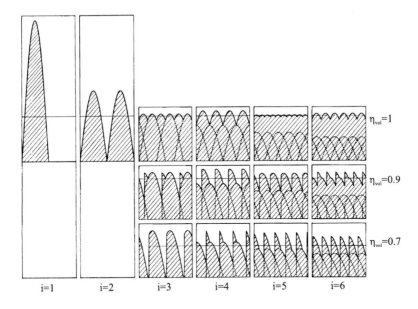

Figure 3.10: Pulsations in the pressure pipe for single and multi piston pumps [47]

3.3 Specific requirements for gasoline high-pressure fuel pumps

Each pump application leads to specific requirements. The main requirement criteria for a high-pressure gasoline pump to be integrated in a common-rail injection system are:

- performance

- integration

- lubrication

- costs

Performance

The commercially available pumps (in 2012) can deliver gasoline at a pressure of up to 20 MPa with a flow rate between $10\,\frac{l}{h}$ and $100\,\frac{l}{h}$ depending on the pump displacement and rotation speed. In the context of this work, a delivery pressure of 80 MPa has been set as an objective. In order to guarantee the flexibility of the common-rail injection system, the flow rate and the delivery pressure have to be independent. All those characteristics are provided by piston pumps [31, 47].

The flow and pressure pulsation should be limited in order to ensure a stable operation of the injection system and prevent any damages in the fuel circuit.

Some companies use triplex pumps, which offer the benefit of reducing significantly the pulsation in comparison to single or dual piston pumps (see subsection 3.2.4). However, most of the recent high-pressure gasoline pumps have a single piston design. This design enables to lubricate the piston driving system (pusher or roller) via the engine oil and to reduce the number of sliding systems as the pump driveshaft is the camshaft of the engine itself. In this case, a higher damping volume should be used, even if it causes a lower reactivity of the injection system [9, 31, 80].

The high-pressure pump cannot be considered as being a wearing part and thus should operate over the whole lifetime of the combustion engine (about 250,000 km) without requiring any maintenance. Its performance (especially its flow rate and delivery pressure) should remain constant over its life-cycle. The piston and cylinder belong to the most critical parts concerned by wear. The wear caused by the reciprocating movement of the piston can increase the gap between the piston and the cylinder, thus increasing the internal leakage and reducing the volumetric efficiency of the pump. As the internal leakage is proportional to the cube of the gap (see equation 3.16), the wear should be limited [61]. Furthermore, a too high wear in the piston driving system (the cam for example) could cause an increase of the dead volume at TDC and contribute to an increase of the volumetric losses (see equation 3.17).

A significant increase of the pump delivery pressure in comparison to commercially available pumps represents an engineering challenge for the reasons shown in figure 3.11. Increasing the delivery pressure by keeping the flow rate constant leads automatically to an increase of the input power. But it also causes higher volumetric (internal leakage, re-expansion in the dead volumes) and mechanical losses (higher friction). In order to compensate the lower volumetric efficiency, the pump displacement has to be increased, thus causing a further increase of the input power and energy losses. Consequently, the lifetime of a pump designed with conventional materials is reduced and should be compensated by the use of advanced materials such as ceramics (see chapter 5).

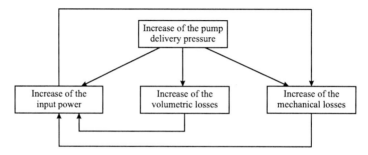

Figure 3.11: Implication of an increase of the pump delivery pressure

Integration

The integration into the environment of the combustion engine is a key point of the high-pressure pump design. The flow rate required by the engine at constant load is proportional to its rotation speed. Since the flow rate of a piston pump is proportional to its speed, driving the pump directly by the combustion engine is the most efficient solution. The mechanical energy needed to drive the pump can be directly used from the engine crankshaft or camshaft. Most of the pumps are driven by one of the camshafts, thus leading to a pump rotation speed which is the half of the engine rotation speed. Furthermore, the pump design should be compact. This characteristic is provided by radial pumps, which are commonly used in modern injection systems [9, 31].

Lubrication

Regardless of what type of seal is used between the piston and the cylinder, the internal leakage cannot be prevented completely. Consequently, the lubrication of the whole pump via the engine oil is not possible. If the oil merges with the fuel flowing through the pump, the hydrocarbon emission in the engine exhaust increases. If the fuel merges with the engine oil, a good engine lubrication is not guaranteed anymore [9]. As a result, the lubrication of the piston/cylinder system or even of the whole gasoline pump is realized by the delivered fuel. This leads to a simple design but also to a significant increase of the tribological stresses in the sliding systems (see section 4.5 of chapter 4).

High-pressure pumps for gasoline should be able to deliver also ethanol as some countries provide fuels with an ethanol content of over 70 %. Even if its viscosity is higher than gasoline and allows a better lubrication of the sliding systems, ethanol implies new challenges regarding material corrosion [53].

Costs

The costs of a high-pressure gasoline pump cannot be neglected. The general design has to be simple by preferring cam/pusher systems with retaining springs instead of crankshaft/rod systems for example. The throttle gap seal between the piston and the cylinder is also preferred to friction sealings. Even if the upper limit of the gap is given by the tolerated internal leakage, the lower limit is given by the operating conditions such as temperature and manufacturing costs. The gap between the piston and the cylinder is a few micrometers [47].

Symbols

Symbol	Unit	Description
α	[°CA]	crank angle
β_T	[Pa^{-1}]	isothermal compressibility of the fluid
ε_0	[-]	compression ratio
φ	[°]	angular phase shift between the pistons
η	[Pa·s]	dynamic viscosity of the fluid

η_{hydr}	[-]	hydraulic efficiency
η_{mech}	[-]	mechanical efficiency
η_{total}	[-]	total efficiency
η_{vol}	[-]	volumetric efficiency
λ	[-]	connecting rod ratio
ρ	[kg/m^3]	fluid density
ω	[rad/s]	angular speed of the driveshaft
c	[m]	clearance between the piston and the cylinder
d	[m]	piston diameter
e	[m]	driveshaft eccentricity
l	[m]	connecting rod length
l_c	[m]	clearance length
$\dot{m}_{real}$	[kg/s]	real massflow
$\dot{m}_{theo}$	[kg/s]	theoretical massflow
M_d	[N·m]	pump driving torque
n	[min^{-1}]	pump rotation speed
p	[Pa]	fluid pressure
p_{high}	[Pa]	high pressure level
p_{in}	[Pa]	pressure in the inlet pipe
p_{low}	[Pa]	low pressure level
p_{max}	[Pa]	maximum cylinder pressure
p_{mi}	[Pa]	indicated mean effective pressure
p_{min}	[Pa]	minimum cylinder pressure
p_{out}	[Pa]	pressure in the outlet pipe
P_{eff}	[W]	pump effective power
P_i	[W]	pump indicated power
P_{in}	[W]	pump input power
r	[m]	crank radius
s_p	m]	maximum piston travel
v	[m/s]	piston speed
V	[m^3]	fluid volume
V_c	[m^3]	compression volume
V_{cyl}	[m^3]	total cylinder volume
V_h	[m^3]	maximum swept volume
V_r	[m^3]	real aspirated volume
$\dot{V}_{real}$	[m^3/s]	real flow rate
$\dot{V}_{theo}$	[m^3/s]	theoretical flowrate
W_i	[J]	pump indicated work
x	[m]	piston travel
z	[-]	number of pistons

4 Tribology

One of the main aspects of the work presented in this document concerns the friction and wear which occur in the sliding systems of gasoline high-pressure pumps (piston/cylinder and piston driving system). These phenomena are part of the interdisciplinary field of tribology, which is described in this chapter.

4.1 Definition and challenges

Tribology can be basically described as a theory about the phenomena resulting from the interaction between two bodies. More precisely, this science is dedicated to the friction occurring in the contact region and to the factors which influence it. Tribology aims at optimizing mechanical systems by reducing the friction losses and wear of the interacting parts. The optimization can affect the performance, the lifetime, the weight, the noise and vibration of a system. Therefore, tribology offers a great potential in the field of science but also in the field of economy since almost one third of the global energy demand is used to compensate friction losses [20, 33].

4.2 Tribotechnical system

The tribotechnical system designates the technical structure which realizes a tribotechnical function by the interaction of surfaces. Such a system can be represented like in figure 4.1. Its structure consists of a base body and a counter body, which are separated by an intermediate and surrounded by an ambient medium. The tribotechnical system is subjected to a stress collective whose main parameters are kinematics, normal force, relative speed between the bodies, temperature and stress duration. The stress collective leads to tribological stresses resulting in a modification of the surfaces as well as energy and material losses. The stress does not only depend on intrinsic properties of the materials but also on several other parameters such as contact geometry, surface structure, surface pressure or lubrication [20, 79, 83].

A tribotechnical system may have an energetic function such as transmitting a force or a movement (bearing, gear), a material function (for example in manufacturing: machining, stamping, etc.), or have an informational function (reading head in a hard-drive for example).

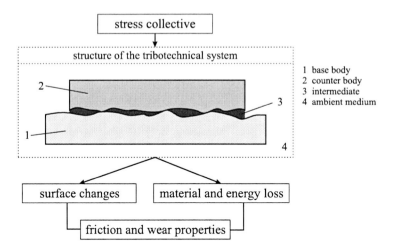

Figure 4.1: Structure of a tribotechnical system [20]

4.2.1 Engineering surfaces

The engineering surface of a body describes its geometrical limit. Its properties have an important influence on the tribological stresses applied to the system. The surface properties of a body can differ from its base material properties. The differences can be explained by various phenomena. On the one hand, the ideal crystal lattice of the base material is interrupted at the surface, thus causing rearrangements due to unsaturated bonds. On the other hand, the material can be subjected to chemical reactions with the ambient medium and cause a modification of the surface composition. Other phenomena can influence the surface structure such as the manufacturing process (machining for example) [20].

Apart from differences in structure or chemical compositions, technical surfaces may also have a different hardness in comparison to the core of the body. This can be explained by the consolidation of the material due to surface deformation. The hardness is a key property to characterize a technical surface as it gives an indication about its elastic limit. Furthermore, the measurement method of the surface hardness is close to the interaction phenomena occurring in the contact during a stress collective. Figure 4.2 presents two common measurement methods of surface hardness: the Vickers (left) and Brinell (right) tests. These two methods are based on the same principle: a hard indenter is pressed in the investigated surface. The area A_h of the impression left by the indenter is compared to the normal force F_n applied and leads to the hardness H of the material via following equation: $H = \frac{F_n}{A_h}$. The measured material hardness is barely influenced by the geometry of the indenter [67].

Another essential property of a technical surface is its roughness. From a microscopic point of view, the surface shows irregularities which differ significantly from the macroscopic geometry. The irregularities consist of peaks and valleys

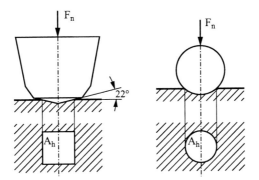

Figure 4.2: Hardness measurement methods: Vickers (left) and Brinell (right) [67]

which are strongly influenced by the manufacturing process. Several parameters can be used to define the roughness of a surface. R_a and R_z belong to the most common parameters used for that purpose and are described in figure 4.3 and by the equations 4.1 and 4.2. The distance between the highest peak and the deepest valley of the investigated surface is given by R_{max} (see figure 4.3 on the top).

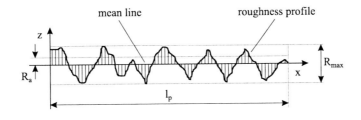

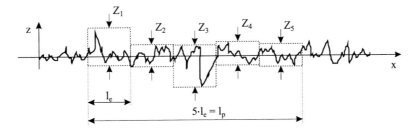

Figure 4.3: Surface roughness measurands [20]

$$R_a = \frac{1}{l_p} \cdot \int_0^{l_p} |z(x)| \, dx \qquad (4.1)$$

$$R_z = \frac{1}{5} \sum_{i=1}^{5} Z_i \qquad (4.2)$$

Another useful characteristic of the microscopic profile is given by the standard deviation of the height of the surface from the center-line σ:

$$\sigma^2 = \frac{1}{l_p} \cdot \int_0^{l_p} z(x)^2 dx \qquad (4.3)$$

In the case of a tribotechnical system consisting in two bodies with different surface roughnesses, the system roughness σ_s is defined as follows:

$$\sigma_s = \sqrt{\sigma_1 + \sigma_2} \qquad (4.4)$$

4.2.2 Contact process

The contact process includes the atomic and molecular interactions as well as the mechanical interactions linked to the geometry and deformation of the surfaces in contact.

Chemical interaction

In a tribotechnical system, the surfaces of the contacting bodies may chemically interact via ionic, covalent, metallic and Van-der-Waals bond forces. This interaction induces the phenomenon of adhesion as the bond forces counteract the relative motion between the contacting bodies [20]. The adhesion is one of the friction types (see section 4.3).

Mechanical interaction

As mentioned in the subsection 4.2.1, the microscopic geometry of a technical surface differs from its macroscopic geometry. The roughness, which can be observed at microscopic scale, shows several micro-contacts which deform under the application of a normal force (see figure 4.4). Consequently, it is necessary to distinguish the geometrical contact area A_{geo} (at macroscopic scale) and the real contact area A_r, which takes the irregularities of the technical surface into account, as defined in equation 4.5.

$$A_{geo} = a \cdot b \gg A_r = \sum_{i=1}^{n} A_r^i \qquad (4.5)$$

The deformation of the surface may be elastic, plastic or viscous-elastic, depending on the material and the contact force. As the materials used for the investigations presented in this document are not viscous-elastic, only elastic and plastic deformation are considered. The real contact area between two curved bodies with perfectly smooth surface has been determined in 1881 by Hertz. His theory has been extended by Archard in 1953 in order to take the rugosity into

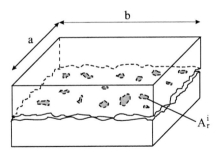

Figure 4.4: Geometrical and real contact area [20]

account. The rugosity is in this case approximated by spherical roughness hills with different diameters. According to this theory, the real contact area A_r is given as a function of the normal force F_n and the elasticity modulus E. Equation 4.6 shows that the real contact area between two bodies of a tribotechnical system is almost proportional to the contact force [20].

$$A_r = const \cdot \left(\frac{F_n}{E}\right)^c \tag{4.6}$$

with $\frac{4}{5} < c < \frac{44}{45}$ depending on the model adopted

4.2.3 Kinematics

The relative motion between the interacting bodies can be of a different nature depending on the considered tribotechnical system. Following relative motions can be distinguished:

- sliding: translation in the contact surface

- rolling: rotation around an "instant axis" on the contact surface

- spinning: rotation perpendicular to the contact surface

- bumping: translation perpendicular to the contact surface with intermittent contact

Depending on the geometry and kinematics considered, the contact surface and the sliding surface may have a different size. This difference is characterized by the contact ratio ε between the contact surface and the total sliding surface [20]:

$$\varepsilon = \frac{\text{contact surface}}{\text{total sliding surface}} \tag{4.7}$$

The contact ratio is a specific property to each body of the tribotechnical system. In the example provided by figure 4.5, the contact ratio ε_1 corresponding to the

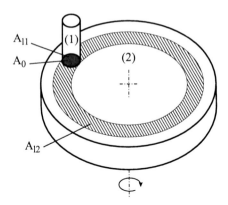

Figure 4.5: Contact ratio in a tribotechnical system [20]

body (1) is given by $\varepsilon_1 = \frac{A_0}{A_{l1}} = 1$. For the body (2), the contact ratio ε_2 is given by: $\varepsilon_2 = \frac{A_0}{A_{l2}} \ll 1$.

A contact ratio equal to one means that the contact and the heat absorption caused by friction are permanent. In addition, the tribochemical reaction between the surface of the body and the ambient medium is limited. In the case of a contact ratio lower than one, the contact is intermittent thus leading to periodic mechanical stress and heat absorption. The surface which is not in contact permanently is more likely to react with the ambient medium.

4.3 Friction

The friction can be defined as the force preventing or counteracting the relative motion between two bodies. Its occurrence is linked to the deformation of the surfaces and to the adhesion phenomenon caused by the chemical interactions in the tribotechnical system. Various types of friction can be distinguished depending on the contact and intermediate medium nature [20]:

- solid body friction (dry friction with direct contact between the friction bodies)

- boundary friction (intensive contact between the surface roughnesses; the friction bodies are coated with a thin lubricant film)

- fluid friction (the bodies are completely separated by a fluid film)

- gas friction (the bodies are completely separated by a gas film)

- mixed friction (the lubricant film is not complete and the surface peaks of the bodies may still come into contact)

4.3.1 Friction mechanism

Depending on the bodies properties, the tribotechnical surfaces and the stress collective, the friction mechanisms likely to occur are listed below (see also figure 4.6):

- adhesion

- deformation

- ploughing

- energy dissipation

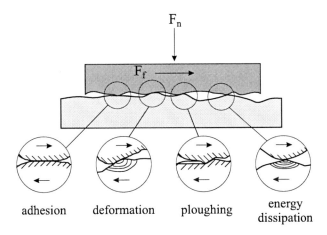

Figure 4.6: Types of friction mechanism [20, 83]

The adhesion mechanism occurs in direct contact regions (area A_r as described in figure 4.4) between the bodies of the tribotechnical system. It is caused by chemical interaction already mentioned in subsection 4.2.2. The adhesion is influenced by the deformation capability of the micro-contact zones and the electronic structure of the materials facilitating bonds between the surfaces. It is also strongly influenced by the boundary layers and the intermediate medium, which can saturate the surface bonds and thus reduce the formation of bonds between the bodies.

The mechanism of contact surface deformation is due to the application of a normal force and the relative motion of the bodies. This mechanism leads to energy losses caused by the plastic deformation of the materials. The deformation is essentially influenced by the stress applied to the tribotechnical system, the material properties and the surface micro-structure.

In the case of surfaces with different hardnesses, the peaks of the hardest body may penetrate the softer body under the application of a normal force. As a result, a relative tangential motion of the bodies may lead to ploughing and cause a resistance to the motion. This mechanism may also appear if hard particles (generated by wear for example) are present in the contact. In the case of brittle materials such as ceramics for example, the ploughing phenomenon may lead to micro cracks in the body. This friction mechanism is mostly influenced by the material properties such as the elasticity modulus, the hardness and the fracture toughness.

In real conditions, the mechanisms listed above may occur simultaneously and interact with each other. They cause energy losses in the form of heat, which is dissipated by the bodies and fluids of the tribotechnical system. The capability of the materials to dissipate heat is given by their thermal conductivity λ [20, 83].

4.3.2 Friction measurands

The friction can be measured by the force counteracting the relative motion between the bodies of a tribotechnical system. This force may be static (no effective relative motion) or dynamic (relative motion between the bodies). In the case of a relative rotation, the local friction forces result in a friction torque [83, 20].

The friction coefficient f provides a characteristic value related to the friction and is given by the ratio of the friction force F_f (tangential to the contact surface) and the contact force F_n (perpendicular to the contact surface):

$$f = \frac{F_f}{F_n} \tag{4.8}$$

Depending on the kinematics of the tribotechnical system, the work of friction W_f is defined as follows:

- sliding friction: $W_f = \int_{s_f} F_f \cdot ds_f$

- rolling friction: $W_f = \int_{\pi_r} M_f \cdot d\pi_r$

- spinning friction: $W_f = \int_{\pi_s} M_f \cdot d\pi_s$

with ds_f, $d\pi_r$ and $d\pi_s$ designating the elementary sliding distance or rotation angle.

The friction can also be characterized by its power by taking the relative speed v_r and the friction duration t into account:

$$P_f = \frac{W_f}{t} = F_f \cdot v_r = f \cdot F_n \cdot v_r \tag{4.9}$$

4.4 Wear

The wear describes the progressive material loss on the surface of the bodies in contact. It can lead to surface damage or to a modification of the bodies geometry and cause a failure of the tribotechnical system. Even if the wear is generally linked to the friction phenomenon, it consists in other mechanisms, which are described in this section [20, 33, 67, 79].

Wear mechanism

Following wear mechanisms may occur in a tribotechnical system (also see figure 4.7):

- surface fatigue

- abrasive wear

- adhesive wear

- wear caused by chemical reactions

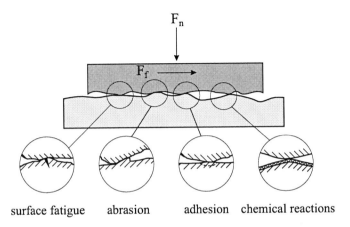

surface fatigue   abrasion   adhesion   chemical reactions

Figure 4.7: Types of wear mechanism [20, 83]

The relative motion between the surfaces of the bodies may lead to a cyclic stress of the surface at microscopic scale. Cyclic stress may induce fatigue and precipitate the failure of the material at the contacting surfaces. Surface fatigue is mainly influenced by the amplitude and the number of stress cycles. This wear mechanism may also cause particle release through cracks formation and propagation.

Abrasion is a phenomenon occurring in the case of one body in contact with a harder and rougher other. This mechanism can also appear if hard particles are present between the contacting surfaces of the tribotechnical system. The peaks of the harder surface penetrate the softer surface through the application of a normal force and remove material through the relative tangential motion of the bodies (micro-ploughing and micro-cutting). This type of wear can be identified by the grooves left on the surface in the relative sliding direction. The abrasive wear amount is mostly influenced by the surface hardness. Hard materials are less subjected to abrasion.

In contrast to the wear mechanisms listed above, adhesion is caused by atomic or molecular scale phenomena. Bonds are created between the bodies through the effect of local forces. A detachment or displacement of the material layer generated by adhesion may occur. This phenomenon is known under the name of seizure in the case of metals. Adhesion is essentially influenced by the crystal structures and electronic structure of the materials in contact.

Chemical reactions may occur between the bodies of a tribotechnical system under the potential influence of intermediate bodies or fluids. The tribological stress applied to the system (relative speed and forces) can also support this mechanism through thermal and mechanical activation. Chemical reactions may alter the properties of the surface layer and lead to an increase or decrease of the wear caused by the other mechanisms listed above [20, 79, 83].

Wear measurands

Wear can be measured in a quantitative and in a qualitative way. Measuring the mass or volume variation of the bodies before and after the stress collective provides a quantitative wear evaluation. The specific wear rate k is a commonly used wear measurand. It is given as a function of the volume lost by wear V_w, the normal force F_n and the cumulative relative sliding distance s_f between the two bodies of the tribotechnical system [20]:

$$k = \frac{V_w}{F_n \cdot s_f} \tag{4.10}$$

Figure 4.8 provides some examples of measured wear rates for various tribotechnical systems, with values in the range of $10^{-4} \frac{\text{mm}^3}{\text{N·m}}$ to $10^2 \frac{\text{mm}^3}{\text{N·m}}$.

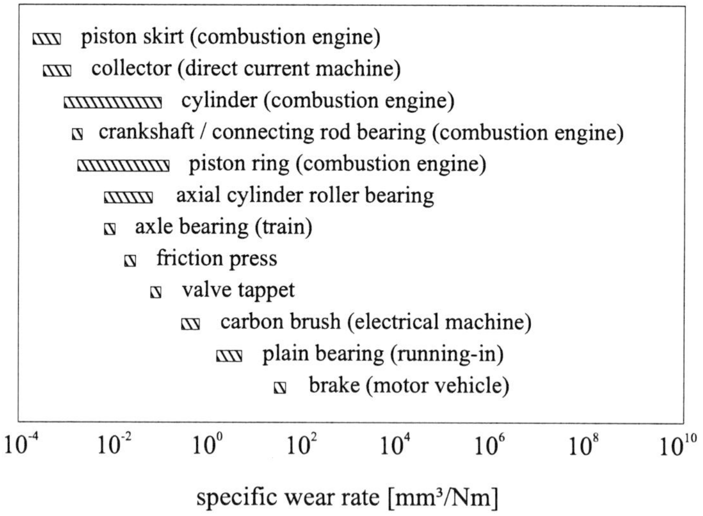

Figure 4.8: Specific wear rate of various tribotechnical systems [79]

The specific wear rate does not provide any indication regarding the friction mechanisms occurring in the tribotechnical system. In addition, it is not specific to any material as it depends on the structure of the system. Therefore, a qualitative analysis (with a microscope for example) of the stressed surfaces or particles generated may help to better understand the origin of wear: depending on the modifications observed between the surfaces before and after the stress collective (grooves, cracks, material transfer, etc.), the friction mechanisms can be identified.

4.5 Lubrication

Lubricating a tribotechnical system is a way to keep the friction forces and wear at a low level and thus to increase the performance and reliability of the system. The lubricant prevents direct contact between the two bodies by generating a film between them. Furthermore, the lubricant enables to better dissipate the heat generated by friction. It may also support chemical reactions in the contact in order to modify the boundary layers and improve their mechanical properties. The lubricant is most of the time liquid (oil for example) but can also be solid or gaseous in some particular cases. In this document, only liquid lubricants are considered.

Lubrication state
Depending on the geometry of the body, the rugosity of its surface, the viscosity of the lubricant and the parameters of the stress collective (relative speed, normal force, temperature, etc.), various lubrication (or friction) states can be observed. The Stribeck curve (see figure 4.9) shows the different states as a function of fluid viscosity η, the relative speed between the contacting surfaces v_r and the normal force F_n [20]. All these parameters affect the lubricant film thickness l_t in the contact: the film thickness increases with increasing viscosity and relative speed but decreases with a higher normal force.

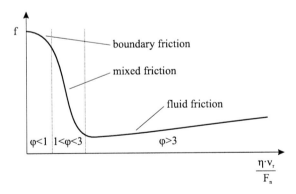

Figure 4.9: Stribeck curve describing the friction or lubrication state of the tribotechnical system

The ratio $\varphi = \frac{l_t}{\sigma_s}$ between the lubricant film thickness and the system roughness σ_s has a direct influence on the lubrication state of the tribotechnical system. If φ tends toward zero, the bodies are in direct contact. This state is called dry friction.

However, the surfaces are generally covered by an adsorption layer consisting of lubricant molecules. This is called boundary friction. For $1 < \varphi < 3$, the friction coefficient decreases significantly. This range is called mixed friction as a part of the friction is directly supported by the peaks of the surfaces whereas another part is supported by the lubricant. For $\varphi > 3$, the friction is entirely supported by the lubricant. Therefore this state is called fluid friction. In this range, the friction coefficient tends to increase with increasing viscosity, normal force and relative speed because of the internal friction in the lubricant. Table 4.1 provides the friction coefficient range of each friction state, with a value between 0.001 in the case of fluid friction and more than 1 in the case of dry friction. In the case of rolling friction, only mixed friction occurs and the coefficient is generally in a range between 0.001 and 0.005.

Friction type	Friction state	Friction coefficient		
Sliding friction	Dry friction	0.1	...	>1
	Boundary friction	0.01	...	0.2
	Mixed friction	0.01	...	0.1
	Fluid friction	0.001	...	0.01
Rolling friction	Mixed friction	0.001	...	0.005

Table 4.1: Friction coefficient range depending on the friction type and state [20]

Gasoline as lubricant

The parameters influencing the lubrication state as described by the Stribeck curve (figure 4.9) enable to better understand the difficulties linked to an increase of the delivery pressure of fuel-lubricated high-pressure gasoline pumps. The low viscosity of gasoline (up to five times lower than Diesel fuel and two times lower than ethanol, see table 4.2), combined to an increase of the contact force in the tribotechnical systems leads to higher friction coefficients, thus reducing the pump mechanical efficiency and reliability. In comparison, modern fuel-lubricated Diesel pumps for automotive application reach a delivery pressure of up to 200 MPa and more [10, 41]. Apart from its viscosity, another factor influencing the lubricity of gasoline is its chemical composition. As previously mentioned, tribochemical reactions may occur in the contact area, modify the material properties on the surface and potentially improve lubrication. But the evolution of the fuel composition, especially the decrease of sulfur content (in order to increase the performance and lifetime of catalysts in the exhaust system) is linked to a decrease of their lubrication properties and may cause adhesion wear [31, 93].

	Gasoline	Ethanol	Diesel
Dynamic viscosity at 20°C [mPa·s]	0.65	1.2	1.2 to 3

Table 4.2: Dynamic viscosity of gasoline, Diesel and ethanol fuels [52, 61, 93]

Symbols

Symbol	Unit	Description
β	[-]	mean diameter of the roughness peaks
ε	[-]	contact ratio of the body
η	[Pa·s]	dynamic viscosity of the fluid
λ	[W/m·K]	thermal conductivity
μ_a	[-]	coefficient of adhesion
ν	[-]	Poisson's ratio
π_r	[°]	rolling angle
π_s	[°]	spinning angle
σ	[-]	standard deviation of the roughness peaks distribution
φ	[-]	lubricant film thickness to roughness ratio
A_{geo}	[m²]	geometrical contact area
A_r	[m²]	real contact area
A_h	[m²]	area of the mark left by the indenter
E	[N/m²]	elasticity modulus
E_f	[J]	friction energy
f	[-]	friction coefficient
F_a	[N]	adhesion force
F_f	[N]	friction force
F_n	[N]	normal force
H	[kg/mm²]	surface hardness
k	[m³/N·m]	specific wear rate
l_e	[m]	sampling length
l_p	[m]	measured profile length
l_t	[m]	lubricant film thickness
M_f	[N·m]	moment of friction
s_f	[m]	sliding (rolling) distance
R_a	[m]	average roughness
R_{max}	[m]	distance between the highest peak and lowest valley in the measured profile length
R_z	[m]	average distance between the highest peak and lowest valley in each sampling length
t	[s]	time (duration)
v_r	[m/s]	relative speed between the friction bodies
V_w	[m³]	wear volume
Z	[m]	distance between the highest peak and lowest valley in the sampling length

5 Advanced ceramics

Ceramic materials represent an alternative to conventional materials. Since they provide excellent properties regarding hardness and compressive strength, the use of ceramic components in the sliding systems of high-pressure pumps should help to increase the injection pressure in GDI engines. This chapter gives the basics concerning advanced ceramics for a better understanding of the work presented in this document.

5.1 Basics

5.1.1 Definition and general properties

Ceramics is a group of non-metallic and non-organic materials, which are manufactured from powder and compacted via a high-temperature process. This general definition includes the traditional ceramics used since more than 25,000 years as well as the advanced ceramics, which have been developed in the last hundred years (see figure 5.1). The traditional ceramics are generally manufactured from clay and silica and are used as bricks or pottery for example. Advanced ceramics provide superior mechanical, thermal, electrical, magnetic, optical or chemical properties, thus offering a large potential of application. They can be used for their functional properties (bio-ceramics, electrical ceramics, sensors, etc.) or their structural properties (material with high strength, for example) [19, 20, 42, 71, 85].

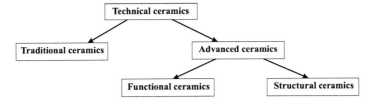

Figure 5.1: Categories of ceramics [71]

Advanced ceramics consist of metal, metalloid and non-metal elements with a combination of covalent, ionic and metallic bonds. They are generally heterogeneous and consist of grains separated by a glass phase. The structure of the ceramic influences its properties significantly. The electronic structure confers optical, electrical or magnetic properties to the material and has an impact on its crystal structure. The size, shape and orientation of the grains affect the macroscopic properties such as the mechanical strength. The properties generally associated to advanced ceramics are their hardness, compressive strength, high-temperature resistance and chemical insensitivity, which can be attributed to the covalent and ionic bonds. However, advanced ceramics are often considered as providing a low tensile and bending strength and having a brittle behaviour at ambient temperature in comparison to metallic materials. The brittleness of ceramics can be explained by their lack of dislocations to relieve stress concentrations. As a result, stress concentration at microscopic failures may cause crack formation and crack growth leading to complete material failure [19, 42, 85].

5.1.2 Types of advanced ceramics and their application

Depending on their composition, advanced ceramics can be classified into one of the three following groups: the silicate ceramics, the oxide ceramics and the non-oxide ceramics. The silicate ceramics are the oldest group and consist in a simple manufacturing process, relatively low sintering temperature and low-cost base materials. These ceramics are based on silicate, which is contained in clay or kaolin for example. Materials such as porcelain or soapstone belong to the silicate ceramics group. The oxide ceramics group includes all the ceramics that are generally composed of metallic oxides and have a very low proportion of glass phase, such as alumina (Al_2O_3), magnesia (MgO) or zirconia (ZrO_2). Those ceramics are manufactured from synthetic materials with a high purity grade. The non-oxide ceramics group designates the ceramics which are based on boron, carbon, nitrogen and silicon bonds. This group includes for example the silicon carbide (SiC), the silicon nitride (Si_3N_4) and the aluminum nitride (AlN) [42].

Ceramic materials can be used in various technical fields as shown in figure 5.2. Structural ceramics such as silicon carbide, silicon nitride or alumina are mostly employed for their mechanical, thermal or chemical characteristics. These materials offer very good properties regarding hardness, mechanical strength (even at high temperatures), wear resistance and corrosion. Structural ceramics are used for many applications such as heat exchangers, cutting tools, and catalysts. Other ceramics are used for their electrical or magnetic properties in electronics (capacities, resistors) or in supra-conductors for example. Barium titanate ($BaTiO_3$), zinc oxide (ZnO) and aluminum nitride provide such properties. Ceramics such as alumina or spinel are employed in lighting application or optics. Inert, bioactive or resorbable ceramics are used for biological applications (human bones replacement for example). Some ceramic types are utilized in nuclear applications as neutron absorbers or reflectors for example [19, 71].

The properties of advanced ceramics can be exploited by the use of massive parts, coatings or films. Ceramics can also be employed in composite materials as matrix phase or reinforcement phase in order to combine the properties of ceramics to the properties of other materials [19].

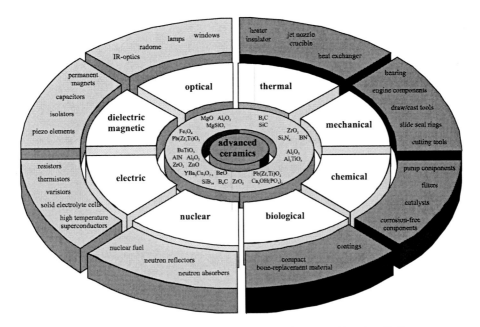

Figure 5.2: Advanced ceramics and their application [71]

5.1.3 Manufacturing process

Figure 5.3 shows the main steps of the ceramic parts manufacturing. It begins with raw natural or synthetic materials in powder form. Additives may be mixed to the powder in order to facilitate the shaping or the sintering. The raw material is prepared in form of a suspension (for casting), granulate or ductile mass (for pressing or extruding). The process of shaping provides a solid form to the material (called green body), which still can be easily machined if necessary. The green body can be shaped by dry or wet pressing, at ambient or high temperature. Other methods such as extrusion or molding are an alternative method of shaping. The green body is usually wet and contains some additives, which have to be eliminated before the next manufacturing steps. Consequently, the green body is dried and the additives are burned out before sintering. The ceramic part is finally obtained by high-temperature sintering (mostly over 1,200 °C). During this step, the particles join together and the porosity decreases. After sintering, the part may only be machined with diamond or ceramic tools as it reached its final hardness [19, 42, 71, 85].

5.1.4 Specific mechanical properties of ceramics

The mechanical properties of ceramic materials can be described by parameters such as their density, their hardness and their bending strength. However, their particularities such as low ductility, low fracture toughness and almost non-existing

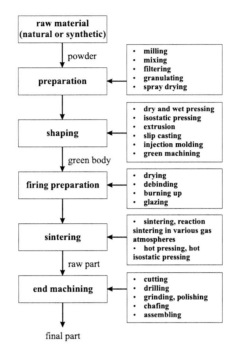

Figure 5.3: Ceramic manufacturing process [42]

plastic deformation should be taken into account since they could lead to a part failure. As a result, the fracture toughness and the probability of failure are two important characteristics of ceramic materials [19].

Fracture toughness

One of the main mechanical properties of a material is its strength. The strength of ceramics strongly depends on the flaws in the material such as pores or micro-cracks, which may propagate. A crack grows if the intensity factor K_I is higher than the critical intensity factor K_{Ic} (also called fracture toughness). K_I is given as a function of the applied stress σ in the crack region, the crack size a and its shape factor Y_I by following equation [19, 71, 85]:

$$K_I = \sigma \cdot \sqrt{\pi \cdot a} \cdot Y_I \tag{5.1}$$

The critical stress intensity factor K_{Ic} provides an information concerning the crack propagation resistance, which directly affects the lifetime of a ceramic part. It can be measured by performing a bending test of a notched sample. The value of the stress intensity factor can be deduced from the applied stress, the sample geometry and the initial crack size. K_{Ic} can also be measured by indentation with a procedure similar to the hardness measurement described in figure 4.2. The size of the cracks initiated by the penetration of the indenter enables to calculate K_{Ic} as a function of the applied stress.

Probability of failure

Ceramic parts are characterized by the scattering of their strength. The resistance of a part with a volume V and subjected to a stress σ is statistically given by a Weibull distribution. The failure probability P_f or the survival probability P_s can be estimated via following equation:

$$P_s = 1 - P_f = \exp\left[-\int_V \left(\frac{\sigma - \sigma_{min}}{\sigma_0}\right)^m dV\right] \tag{5.2}$$

σ_{min} designates the stress level below which the probability of failure is zero. σ_0 indicates the stress level at which the probability of failure is $1/e \simeq 0.37$. The coefficient m is the Weibull modulus. Its value depends on the material (between 5 and 20 in the case of ceramics). A higher Weibull modulus leads to a lower scattering of the material strength. Equation 5.2 shows that the probability of failure increases with increasing part volume. Figure 5.4 shows the evolution of the survival probability of a ceramic part as a function of the applied stress σ according to the equation 5.2 for different Weibull modulus values [19, 71].

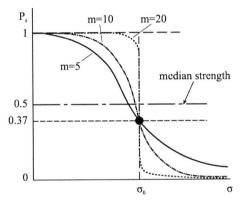

Figure 5.4: Survival probability given by the Weibull distribution [19]

5.1.5 Design of ceramic components

The specific properties of ceramic materials listed in this section have an influence on the part design: it differs significantly from metallic materials because of their different manufacturing process and mechanical properties. As ceramics provide a tensile strength which is lower than their compressive strength, designing ceramic parts under compressive stress only is preferable. If combined stress (tensile and compressive) cannot be avoided, the ceramic components should be pre-loaded by shrinking them into metallic parts for example. This method enables to compensate bending or tensile stresses. The stress concentrations should also be avoided as they could initiate crack growth and lead to a part failure. Consequently, the stress should be distributed homogeneously on the surface and sharp edges should be prevented. The survival probability of a ceramic part can be maximized by reducing its volume, as can be inferred from equation 5.2. By reducing the volume of a part, the presence probability of critical failures decreases. Finally, machining sintered parts is difficult because of the high hardness of the material. As a result, simple shapes requiring minimal post-sintering machining and large tolerances (taking the deformation occurring during the sintering process into account) should be preferred [42, 85].

Symbols

Symbol	Unit	Description
σ	[MPa]	applied stress
σ_0	[MPa]	stress at which the stress probability is $1/e$
σ_{min}	[MPa]	stress below which the failure probability is 0
a	[m]	crack size
K_I	[MPa·m$^{1/2}$]	stress intensity factor
K_{Ic}	[MPa·m$^{1/2}$]	critical stress intensity factor (fracture toughness)
m	[-]	Weibull modulus
P_f	[-]	probability of failure
P_s	[-]	probability of survival
V	[m^3]	stressed volume
Y_I	[-]	crack shape factor

6 Investigations

Two prototype pumps have been used in order to demonstrate the potential of ceramic sliding systems for high-pressure gasoline pumps. A single-piston pump delivering fuel at up to 50 MPa has been tested in order to measure the friction coefficients in the main sliding systems (piston/cylinder and cam/pusher). The objective was to determine which material pairs and which surface finish best fit an application in a high-pressure gasoline pump regarding friction and wear. Based on the knowledge accumulated with the single-piston pump, a 3-piston-pump has been designed. The maximum delivery pressure of the 3-piston pump has been set to 80 MPa. This pressure level is a compromise between the advantages of high-pressure injection and the increased driving power required by the pump [15, 75]. The 3-piston prototype pump is only lubricated by the delivered fuel and its design is more production-oriented than the single-piston pump. It enables to measure the total pump efficiency in realistic operating conditions. Furthermore, two commercially available pumps have been tested in order to provide a reference for the prototype pumps. This chapter describes the investigated pumps, the test bench and the testing conditions, which led to the results provided in the next chapter.

6.1 Investigated high-pressure fuel pumps

6.1.1 Single-piston prototype pump

The single-piston pump used for the investigations is based on the design of a 3-piston radial pump, which is considered as being the best compromise between size, efficiency and costs [9]. This prototype pump results from a constant development since the year 2000 within the Collaborative Research Centre SFB483. The pump includes several sensors such as force, pressure and temperature transducers in order to assess the performance of the materials used in the piston/cylinder and cam/pusher systems. Both systems are only lubricated by the delivered fuel.

Design and kinematics
The kinematics of the pump is shown in figure 6.1. As the eccentric shaft rotates, the motion of the cam is described as a circular translation. The cam drives the piston via the pusher. The piston/pusher assembly maintains contact with the cam via a retaining spring, which is not shown in the figure. The sealing of the piston/cylinder system is realized by a throttle gap seal (see subsection 3.2.3). This design causes rocking of the piston in the cylinder at approximately 90°CA and 270°CA after bottom dead centre (BDC), since the sliding sense of the cam/pusher

system is reversed at these shaft angles. Consequently, the piston is in contact with the cylinder at only two points: B and C or B^* and C^* (see figure 6.3). The joints between the driveshaft and the eccentric lobe and between the driveshaft and the body are lubricated by oil. The position and the relative sliding speed between the piston and the cylinder are given by the equations 3.2 and 3.5 respectively. The relative sliding speed in the cam/pusher system is identical to the piston/cylinder system with a phase shift of 90 °CA.

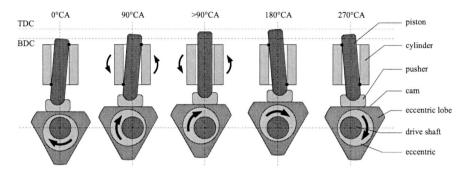

Figure 6.1: Kinematics of investigated high-pressure pumps

The lubrication in the piston/cylinder and cam/pusher systems is provided by the fuel leaking out of the work chamber. However, as the leak flow may not guarantee sufficient lubrication, an auxiliary lubrication system has been added. It consists of a small pipe connected to the low-pressure circuit (see figure 6.10) and supplying fuel directly in the cam/pusher system.

Geometry of the investigated components

The piston and the cylinder (see figure 6.2) both have a nominal diameter of 12 mm. The maximum piston travel is 6 mm (the driveshaft has an eccentricity of 3 mm), for a theoretical displaced volume of 678 mm^3. The length of the cylinder is 25 mm. In order to limit leakage while allowing satisfactory lubrication of the sliding contact, the gap between the piston and the cylinder has been set to 7 μm. A metal jacket is glued to the pistons made of ceramic for easier integration in the pump. The cylinder is shrunken into a metallic jacket in order to keep the ceramic under compression stress. The cam and the pusher both consist of a ceramic part glued into a metallic holder, which facilitates the replacement of the material pairs in the pump. The cam and the pusher have an external diameter of 49 mm and 40 mm, respectively. The amplitude of their relative motion is 6 mm (linked to the eccentricity of the driveshaft). The pusher is ring-shaped with an internal diameter of 28 mm in order to keep more fuel in the contact area, thus supporting a better lubrication. Both surfaces are flat and there is a light chamfer (0.5 mm, 15°) on the ring. Taking the dimensions of the components, the maximum delivery pressure (50 MPa) and the retaining spring force into account, the maximum surface pressure in the cam/pusher system is approximately 11 MPa.

According to the geometry of the contact surfaces, the contact ratio of the cam as defined in subsection 4.2.3 is $\varepsilon_{cam} = 0.57$. The contact ratio of the pusher is $\varepsilon_{pusher} = 1$ as its whole surface is always in contact with the cam.

Figure 6.2: From left to right: piston, cylinder, cam and pusher in the single piston pump

Sensors

Figure 6.3 provides a perspective view of the single-piston prototype pump and a schematic view of the force measuring principle. The value supplied by the cam force sensor indicates the friction force in the cam/pusher system. The contact force in this system can be calculated by summing the force applied to the piston by the fuel pressure in the cylinder, the value supplied by the z force sensor and the force applied by the retaining spring (proportional to the piston travel). The z sensor directly measures the friction force in the piston/cylinder system since a second piston compensates the pressure force in the z direction (pressure compensation piston in figure 6.3). As the cylinder assembly is articulated at A, it is possible to calculate the contact force applied in B and C (respectively B^* and C^*) with the value supplied by x and y force sensors. More details about the forces calculation method are given in [37].

A temperature sensor located in the metallic cam holder provides information about the heat release level in the cam/pusher system in order to prevent any insufficient lubrication and cooling during the experiments. An angle sensor with a resolution of 1° CA provides the angular position of the pump driveshaft. Table 6.1 shows the type of sensor that were used in the single-piston pump and their measuring range.

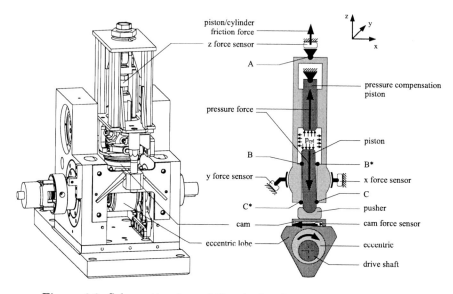

Figure 6.3: Schematic view of the single-piston prototype pump

Sensor	Measuring principle	Measuring range	Manufacturer / Type
cylinder pressure	piezoresistive	0..100 MPa	Kistler / 4067 A1000 A2
force in x-direction	piezoelectric	±5 kN	Kistler / 9311 B
force in y-direction	piezoelectric	0..14 kN	Kistler / 9133 B21
force in z-direction	piezoelectric	±10 kN	Kistler / 9321 B
cam force	piez4oelectric	0..15 kN	Kistler / 9011 A
driveshaft angle	optical	1°	Heidenhain / ROD 426 360
cam temperature	thermocouple	-200..+1150 °C	Electronic Sensor / IKL 10/25

Table 6.1: Sensors used in the single-piston pump

6.1.2 3-piston prototype pump

Design and kinematics

The 3-piston pump is a radial pump with an angular shift of $\varphi = 120\,°\text{CA}$ between the pistons. It has been designed to deliver fuel at a pressure of up to 80 MPa. Its kinematics is similar to the single-piston pump (see figure 6.1). The piston/cylinder, cam/pusher and driveshaft/eccentric lobe are lubricated by the delivered fuel, which flows through the housing before being suctioned into the cylinders. The joint between the driveshaft and the pump housing is realized by grease lubricated ball bearings. Figure 6.4 depicts the main components of the 3-piston prototype pump.

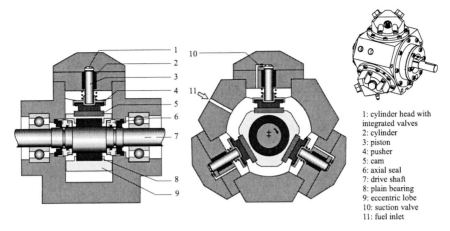

1: cylinder head with
integrated valves
2: cylinder
3: piston
4: pusher
5: cam
6: axial seal
7: drive shaft
8: plain bearing
9: eccentric lobe
10: suction valve
11: fuel inlet

Figure 6.4: Schematic view of the 3-piston prototype pump

Geometry of the investigated components

The 3-piston pump differs from the single-piston pump in the geometry of the components used in the sliding systems (see figure 6.5). The nominal diameter of the pistons is 8 mm and the piston travel is 4 mm. The cylinder length has been set to 16 mm. The total pump displacement is 603 mm^3. As the dimensions have been reduced in comparison to the single-piston prototype pump, the gap between the cylinders and the pistons has been reduced to 5 µm in order to limit the pump volumetric losses. The cams and the pushers have a circular shape with a diameter of 25 mm and 20 mm, respectively. There is a light chamfer (0.5 mm, 45°) on the edges of the cams and pushers. At 80 MPa delivery pressure, the surface pressure in the cam/pusher systems is approximately 14 MPa (corresponding to a ratio of 0.175 between the surface pressure in the contact and the pump delivery pressure). Taking the geometry of the contact surfaces in the cam/pusher system into account, the contact ratios of the cam and pusher are $\varepsilon_{cam} = 0.79$ and $\varepsilon_{pusher} = 1$, respectively.

The joint between the eccentric shaft and the eccentric lobe is realized by a polished plain bearing made of silicon carbide (EKasic F from ESK Ceramics). In contrast to the piston/cylinder and cam/pusher systems, the relative sliding speed between the eccentric shaft and the eccentric lobe remains constant over time at constant pump rotation speed, thus favoring the lubrication of the system. Therefore, this system is not investigated in this document.

Sensors

The 3-piston prototype pump is fitted with pressure sensors in each cylinder and a temperature sensor is located in the pump housing. These sensors are the same type as those listed in table 6.1. In contrast to the single-piston pump, the 3-piston pump is not fitted with force transducers. Only the resulting effect of the internal pressure and friction forces in the pump can be measured via a torque sensor.

Figure 6.5: From left to right: piston, cylinder, cam and pusher in the 3-piston prototype pump

Balance of forces

Figure 6.6 depicts the balance of forces applied to three different groups in the 3-piston prototype pump. This description osee section 6.4)f the internal forces enables to better understand the calculation of the theoretical driving torque mentioned in the next chapter.

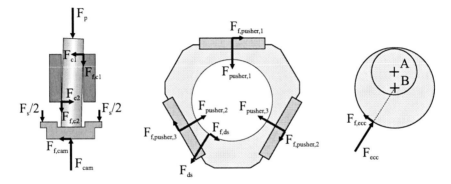

Figure 6.6: Balance of forces applied to the piston group (left), eccentric lobe (middle) and driveshaft (right) in the 3-piston prototype pump

The draft on the left of the figure shows the balance of forces applied to the piston group (including the pusher). The force applied by the cam to the pusher is a function of the pressure force F_p, the force applied by the retaining spring F_s and the friction forces between the piston and the cylinder ($F_{f,c1}$ and $F_{f,c2}$):

$$\overrightarrow{F_{cam}} = -\overrightarrow{F_p} - \overrightarrow{F_{f,c1}} - \overrightarrow{F_{f,c2}} - \overrightarrow{F_s} \tag{6.1}$$

The friction force $F_{f,cam}$ between the cam and the pusher causes the contact forces F_{c1} and F_{c2} between the piston and the cylinder:

$$\overrightarrow{F_{f,cam}} = \overrightarrow{F_{c1}} - \overrightarrow{F_{c2}} \tag{6.2}$$

In the middle of figure 6.6 is shown the balance of the forces applied to the eccentric lobe group (including the cams). This group is subjected to the contact and friction forces from the three piston groups ($F_{pusher,i} = -F_{cam}$ and

56

$F_{f,pusher,i} = -F_{f,cam}$) and from the eccentric shaft (F_{ds} and $F_{f,ds}$). These forces are linked via following equation:

$$\vec{F_{ds}} + \vec{F_{f,ds}} + \sum_{i=1}^{3}(\vec{F_{pusher,i}} + \vec{F_{f,pusher,i}}) = \vec{0} \qquad (6.3)$$

The forces applied to the eccentric shaft are depicted on the right of figure 6.6. The direction of the contact force $F_{ecc} = -F_{ds}$ applied to the eccentric lobe crosses the center of the eccentric shaft (point B in the figure). The relative motion between the eccentric lobe and the eccentric shaft leads to the friction force $F_{f,ecc} = -F_{f,ds}$. Both contact and friction forces are linked to the driving torque T_d applied to the driveshaft in point A:

$$\vec{T_d} = (\vec{F_{ecc}} + \vec{F_{f,ecc}}) \wedge \vec{AB} \qquad (6.4)$$

The friction coefficients in the cam/pusher, piston/cylinder and eccentric lobe/eccentric shaft systems are defined by following equations:

$$f_{cam/pusher} = \frac{F_{f,cam}}{F_{cam}} \qquad (6.5)$$

$$f_{piston/cylinder} = \frac{F_{f,c1} + F_{f,c2}}{F_{c1} + F_{c2}} \qquad (6.6)$$

$$f_{lobe/shaft} = \frac{F_{f,ds}}{F_{ds}} \qquad (6.7)$$

6.1.3 Commercially available high-pressure pumps

Two production pumps have been investigated in order to better evaluate the performance of the prototype pumps. The one is a high-pressure gasoline pump (Bosch HDP1) whereas the other is a high-pressure Diesel pump (Bosch CP1). Both pumps have been chosen because of their 3-piston radial design, which is similar to the prototype pump designed for the experiments presented in this document.

HDP1
The HDP1 is a high-pressure gasoline pump used in the first generation of GDI engines. It provides a nominal fuel pressure of 12 MPa and a maximum fuel pressure of 20 MPa for short time intervals. The cam/pusher, piston/cylinder and the plain bearings are lubricated by the delivered fuel. The kinematics of the HDP1 is similar to the prototype pumps previously described and is based on a cam/pusher system. The driveshaft is guided by an oil-lubricated ball bearing at one side and by a fuel-lubricated plain bearing at the other side. The joint between the eccentric lobe and the eccentric shaft is realized by a fuel-lubricated plain bearing. One particular feature of this pump is that the suction valves are integrated into the pistons (see figure 6.7). The displacement of the investigated

HDP1 is $400\,mm^3$ (piston diameter: $6.5\,mm$; piston travel: $4\,mm$). The pusher has a diameter of $15\,mm$, thus leading to a surface pressure of $3.8\,MPa$ in the cam/pusher system at $20\,MPa$ delivery pressure. The ratio between the surface pressure in the cam/pusher system and the delivery pressure is 0.188. Five grooves in the surface of the cam ($1\,mm$ width, $3\,mm$ between each groove) support the lubrication and cooling of the sliding system. The contact ratio of the cam and the pusher is $\varepsilon_{cam} = 0.74$ and $\varepsilon_{pusher} = 0.62$, respectively. Figure 6.7 provides a sectional view of the Bosch HDP1 with its main components [31].

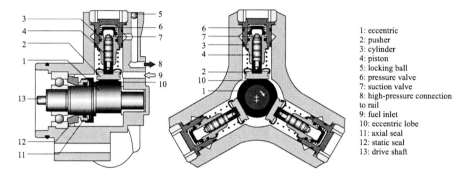

Figure 6.7: Sectional view of the Bosch HDP1 [31]

CP1

The Bosch CP1 has been selected for the experiments as its design is very close to the design of the Bosch HDP1. The CP1 is capable of delivering Diesel fuel at up to $135\,MPa$. It is lubricated by the delivered fuel only. Its kinematics is identical to the other pumps presented in this document with pistons driven by cam/pusher systems. The driveshaft is guided by two fuel-lubricated plain bearings. The displacement of the CP1 used for investigations is $700\,mm^3$ (piston diameter: $6.5\,mm$; piston travel: $7\,mm$). The diameter of the pusher is $16\,mm$. The ratio between the surface pressure in the cam/pusher system and the delivery pressure is 0.175. The contact ratios of the cam and pusher is $\varepsilon_{cam} = 0.64$ and $\varepsilon_{pusher} = 1$, respectively. Figure 6.8 provides a sectional view of the Bosch CP1 and its main components [30].

6.2 Material tested in the sliding systems

6.2.1 Ceramic materials

The materials used in the tribotechnical systems of a fuel-lubricated high-pressure gasoline pump should have specific properties such as a high hardness, wear resistance and corrosion resistance. Consequently, following ceramics were selected for the investigations presented in this work: silicon carbide, silicon nitride and sialon.

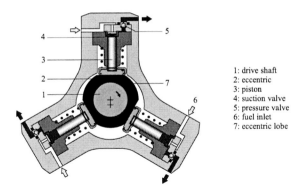

1: drive shaft
2: eccentric
3: piston
4: suction valve
5: pressure valve
6: fuel inlet
7: eccentric lobe

Figure 6.8: Sectional view of the Bosch CP1 [30]

Silicon carbide (SiC)

Silicon carbide has a cubic crystal structure (β-SiC) or a hexagonal crystal structure (α-SiC). The β phase transforms into α phase at approximately 2,100 °C, thus giving to these phases the name of low-temperature and high-temperature phase, respectively. The high ratio of covalent bonds in silicon carbide provides a high hardness but also makes its sintering difficult. Various manufacturing processes (pressure-less sintering, hot pressing, hot isostatic pressing, etc.) can be used in order to influence the properties of the final material. The process selected for the investigations presented in this document is the pressure-less sintering, which enables to reach a density higher than 95 % of the theoretical maximum density, thus meaning a very low porosity. The material obtained via pressure-less sintering is called SSiC (sintered silicon carbide).

The main properties of silicon carbide are its mechanical strength (bending strength of up to 400 MPa at ambient temperature), its hardness (up to 2,700 HV) and its high-temperature operability (1,400 °C and more). It provides a good thermal conductivity, which enables to better dissipate heat generated by friction in a tribotechnical system for example. Silicon carbide also offers a very good resistance to corrosion and can be used in systems under high mechanical, thermal and chemical stress. It is notably used in the design of slide ring sealings or turbines components [8, 57, 71, 85]. Moreover, several investigations have demonstrated that it offers very good performance regarding friction coefficients and wear resistance in tribotechnical systems, which are similar to those found in high-pressure gasoline pumps [38, 98].

Silicon nitride (Si$_3$N$_4$)

Silicon nitride is mainly composed of two phases with a hexagonal and a rhombohedral crystal structure respectively called α-Si$_3$N$_4$ and β-Si$_3$N$_4$. The β-phase offers a better fracture toughness and a better mechanical strength than the α-phase. The α-phase transforms into β-phase at high temperature in the presence of impurities. It appears in the form of elongated grains. Two manufacturing processes are mainly used to produce silicon nitride: nitriding elementary silicon (reaction bonded silicon nitride) or sintering a fine powder of Si_3N_4 by liquid phase sintering. The first process leads to a high porosity (10 % to 30 %) and to a low mechanical strength of the material. The second process can be realized via various methods such as pressure-less sintering, gas pressure sintering, hot pressing or hot isostatic pressing and leads to better properties in comparison to the nitriding process. It is possible to combine the two processes in order to eliminate the disadvantages of nitriding only and thus to improve the mechanical strength of the material. Another method called FAST (Field Assisted Sintering Technique) enables to sinter the green body in much shorter time (few minutes) by using a pulsating direct current flowing through the material as a heat source.

Silicon nitride has a higher density than silicon carbide ($3.2 \frac{g}{cm^3}$ to $3.6 \frac{g}{cm^3}$ for silicon nitride and $2.6 \frac{g}{cm^3}$ to $3.2 \frac{g}{cm^3}$ for silicon carbide) but provides a lower thermal conductivity ($10 \frac{W}{m \cdot K}$ to $30 \frac{W}{m \cdot K}$). Its bending strength is higher than silicon carbide as it can reach 1,200 MPa at ambient temperature. Its fracture toughness K_{Ic} is also better with a value of up to $7 \, MPa \cdot m^{\frac{1}{2}}$ ($4 \, MPa \cdot m^{\frac{1}{2}}$ for silicon carbide). Like silicon carbide, silicon nitride offers a very good wear resistance. It is employed in cutting tools, turbines, valvetrains or bearings for example [8, 39, 40, 57, 71, 85].

Sialon (SiAlON)

Sialon (Si-Al-O-N) is a derivative of silicon nitride with a controlled proportion of silicon and nitrogen replaced by aluminum and oxygen atoms. This solid solution generally appears in two modifications, α- and β-Sialon. The α-phase with a rhombohedral structure has the general chemical notation $Me_{n+1}Si_{12-(n \cdot x+y)} Al_{n \cdot x+y} O_y N_{16-y}$ (Me designates one of the following elements: Ca, Mg, Y, Li, Mn or rare earth) and appears in the form of equiaxed grains. The β-phase with a hexagonal structure has the chemical notation $Si_{6-x}Al_xO_xN_{8-x}$ (with $x < 4.2$) and appears in the form of elongated grains. The α-phase provides a higher hardness and a higher wear resistance whereas the β-phase offers a better fracture toughness and a higher thermal conductivity. Consequently, controlling the proportion of these two phases (through the choice of the raw material and sintering additives) enables to adjust the material properties with a higher flexibility than pure silicon nitride. In order to reach a high density, sialon is sintered in liquid phase at more than 1,400 °C. Using FAST sintering is also a possibility.

By varying the α/β phase ratio, the properties of sialon can be adjusted to the requirements of the application. The β-phase can reach a hardness similar to silicon nitride (approximately 1,700 HV) and a fracture toughness of up to $6 \, MPa \cdot m^{\frac{1}{2}}$ whereas the α-phase can reach a value of more than 2,000 HV with a still good fracture toughness. The bending strength of sialon is higher than silicon carbide with a value of up to 1,000 MPa at ambient temperature. However, its thermal conductivity is similar to silicon nitride. Investigations performed on a tribometer in [2] have shown that the use of sialon in a tribotechnical system may lead to lower friction coefficients and wear levels in comparison to silicon nitride. Furthermore, sialon provides a better resistance to corrosion when compared to silicon nitride [2, 25, 69, 70, 71, 85].

Sialon-silicon carbide composite (SiAlON-SiC)

Silicon carbide, silicon nitride and sialon show great potential for their use in tribotechnical systems like those to be found in a high-pressure gasoline pump (high hardness, good wear resistance, low friction coefficients, resistance to corrosion). Each one provides advantages and disadvantages. For instance, silicon carbide offers a high hardness and a good thermal conductivity but its mechanical strength is lower than silicon nitride or sialon. In addition, silicon carbide is lighter but its fracture toughness is lower in comparison to silicon nitride or sialon.

Producing a composite such as sialon-silicon carbide (SiAlON-SiC) allows to exploit the benefits of each ceramic. Investigations performed in [69] have shown that an increasing ratio of silicon carbide in a sialon matrix leads to better tribological performances. The increased hardness and the higher thermal conductivity of this composite offer lower friction coefficients in comparison to pure sialon. The optimal ratio seems to be reached with 30 % of silicon carbide in the sialon matrix.

Silicon nitride with carbon nanotubes (Si_3N_4 CNT)

In order to improve the tribological performances (friction coefficient, wear) of ceramic materials, a further approach has been explored in [32]. It consists in developing a nano-composite made of silicon nitride and 5.3 % multiwalled carbon nanotubes sintered with FAST method (5 minutes at 1,585 °C, 50 MPa pressure applied, 4 Pa vacuum atmosphere).

Carbon nanotubes have already shown their potential regarding lubrication in polymeric and metallic materials. The addition of carbon nanotubes (CNT) in the silicon nitride matrix leads to a lower elastic modulus, a lower hardness and a lower fracture toughness in comparison to pure silicon nitride. However, investigations on a tribometer (ball on plate, lubrication with isooctane) have shown that adding carbon nanotubes to silicon nitride enables to reduce the friction coefficient and the wear significantly since the nanotubes act as a solid lubricant in the tribotechnical system and better redistribute local stresses [21, 32, 72].

6.2.2 Material combinations

Various material combinations have been tested in the piston/cylinder and cam/pusher systems of the prototype pumps. The investigations presented in this document focus on the cam/pusher system. In the cam/pusher system, the contact force between the two bodies is directly proportional to the cylinder pressure. The performance of this system is essential as it directly affects the performance of the piston/cylinder system. The contact forces between the piston and the cylinder are proportional to the friction forces in the cam/pusher system. As a result, low friction between the cam and the pusher leads to benefits in both investigated sliding systems.

Piston/cylinder

For the reasons previously mentioned, only two material pairs were used in the piston/cylinder system of the single-piston prototype pump: self-mated silicon carbide (SSiC) and silicon carbide combined with hardened bearing steel AISI 52100.

This type of steel is used for camshafts in combustion engines for example [54]. Both material pairs have shown good performance regarding friction coefficients and wear in experiments performed by Wöppermann in [98] and Häntsche in [37]. The 3-piston prototype pump has only been tested with a combination of silicon carbide and AISI 52100 in its piston/cylinder systems.

Cam/pusher

Various material combinations have been tested in the single-piston prototype pump in order to evaluate their performance depending on the mean relative sliding speed and surface pressure in the system. The same material pairs have been tested in the 3-piston pump under more realistic conditions. As the components of the 3-piston prototype pump are easier to manufacture (smaller size, simpler shape), some material pairs were only tested in this prototype.

The materials used as cams are SSiC (eventually textured, see subsection 6.2.4), SiAlON (with a α/β phase ratio of 60/40 and 90/10), SiAlON-SiC composite (based on the same base powder than the pure SiAlON tested in this work) and Si_3N_4 (with a α/β phase ratio of 40/60, eventually containing carbon nanotubes). These materials were self-mated or combined to AISI 52100 in some cases. Table 6.2 lists all the investigated material combinations.

	Pusher material	
Cam material	Same as cam material	AISI 52100
SSiC	single-piston pump 3-piston pump	single-piston pump
SSiC (textured)	single-piston pump 3-piston pump	single-piston pump 3-piston pump
SiAlON (α/β: 60/40)	single-piston pump 3-piston pump	-
SiAlON (α/β: 90/10)	3-piston pump	-
SiAlON-SiC	3-piston pump	-
Si_3N_4 (α/β: 40/60)	3-piston pump	3-piston pump
Si_3N_4 (α/β: 40/60, CNT)	3-piston pump	3-piston pump

Table 6.2: Material combinations investigated in the cam/pusher systems of the prototype pumps

6.2.3 Material properties

Among the investigated materials, some are commercially available whereas others are produced by research institutes. The sintered silicon carbide is the EKasic F from ESK Ceramics. The different types of sialon (including the SiAlON-SiC composite) have been developed and produced at the IAM-KM from the KIT. The sialon has been densified by gas pressure sintering whereas the SiAlON-SiC composite has been sintered by Field Assisted Sintering Technique (FAST). Further information about the different types of sialon investigated in this document are available in [69]. The silicon nitride (including silicon nitride containing carbon nanotubes, written CNT) has been developed and produced at the Institute of Ceramics and Glass (ICV) from Spanish National Research Council (CSIC), Spain. It has been sintered via FAST. More details about the investigated silicon nitride are given in [32]. The AISI 52100 is provided by Eisen Schmitt. In order to increase its hardness, it has been heated to 870 °C during 30 minutes in a vacuum furnace and oil quenched. Then, the parts were annealed at 300 °C during one hour for stress relief. Table 6.3 provides the main properties of the investigated materials.

Property	SSiC	SiAlON (60/40)	SiAlON (90/10)	SiAlON-SiC	Si_3N_4 (40/60)	Si_3N_4 (CNT)	AISI 52100
Density [kg·m^{-3}]	3.15	3.30	3.39	3.33	3.23	3.15	7.84
Elastic modulus [GPa]	410	319	321	345	322	222	210
Bending strength [MPa]	405	904	959	N/A	795	N/A	N/A
Tensile strength [MPa]	N/A	N/A	N/A	N/A	N/A	N/A	620
Weibull modulus [-]	7.2	6.2	12.3	N/A	N/A	N/A	N/A
Fracture toughness [MPa·m$^{1/2}$]	3.65	6.5	6.2	5.6	5.6	4.4	N/A
Hardness (Vickers) [-]	2540 HV0.5	1850 HV10	1970 HV10	2030 HV10	1930 HV5	1270 HV5	790 HV30
Thermal conductivity [W/m·K]	110	20	20	N/A	14.9	10.1	37.0

Table 6.3: Characteristics of the investigated materials [69, 98]

6.2.4 Surface finish

Grinding method and surface roughness
All the samples tested in the cam/pusher systems (both prototype pumps) have been fine ground with a diamond disc D25. This technique leads to a surface roughness in the range of $R_a = 0.1\,\mu m$. Two complementary surface roughnesses have been tested on silicon carbide: polished and rough ground. The pistons and cylinders have been polished in order to reach a surface roughness of $R_a < 0.05\,\mu m$. Table 6.4 gives the measured roughness of the investigated components.

Texture
In order to improve the performance of the sliding systems, the surface of some silicon carbide cams has been textured with a laser. The micro-structure is expected to keep the fuel in the sliding system on the one hand and to store the particles released by wear on the other hand. The texture used for the investigations has been specially developed and realized at the IAM-AWP at the KIT

Material	Roughness R_a [µm]	
	Cam	Pusher
SSiC (fine ground)	0.15	0.17
SSiC (rough ground)	0.20	0.50
SSiC (polished)	0.006	0.006
SiAlON (60/40)	0.06	0.16
AISI 52100	-	0.15

(a) Single-piston prototype pump

Material	Roughness R_a [µm]	
	Cam	Pusher
SSiC (fine ground)	0.08	0.09
SiAlON (60/40)	0.11	0.08
SiAlON (90/10)	0.11	0.08
SiAlON-SiC	0.09	0.11
Si_3N_4	0.10	0.09
Si_3N_4 CNT	0.11	0.09
AISI 52100	-	0.05

(b) 3-piston prototype pump

Table 6.4: Surface roughness of the investigated cams and pushers

for the present application [98]. This texture is shown in figure 6.9 and has the following characteristics:

- shape: circular micro-dimples

- diameter: 60 µm

- depth: 10 µm

- surface ratio: 20 %

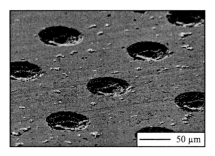

Figure 6.9: SEM-picture of a textured cam made of silicon carbide

6.3 Fuels

A first test series has been performed on the single-piston prototype pump with isooctane as delivered fuel and lubricant. Using isooctane enables to avoid the influence of fluctuations in gasoline composition. Complementary tests have been performed with commercial gasoline (RON 95, containing 5 % ethanol) in order to ensure the transferability of the results. As the proportion of ethanol in commercial fuels is meant to increase in the future due to new legislation, ethanol (E99, containing 1 % methyl ethyl ketone) and gasoline-ethanol mixtures (E75 and E25 with 75 % and 25 % ethanol, respectively) as delivered fuel and lubricant have also been used. The 3-piston prototype pump has been tested with commercial gasoline (RON 95, containing 5 % ethanol) and ethanol (E99, containing 1 % methyl ethyl ketone) only. Table 6.5 shows the density, dynamic viscosity and boiling point of the fuels used for the experiments.

	Isooctane	Gasoline	Ethanol	E75	E25
Density [kg·m^{-3}]	692	750	794	779	750
Dynamic viscosity [mPa·s]	0.348	0.65	1.2	N/A	N/A
Boiling point [°C]	99	40..204	78	47..171	34..198

Table 6.5: Main characteristics of the fuels used for the investigations

6.4 Test bench

The investigated pump is driven by a variable speed asynchronous motor. The pump is integrated in a hydraulic system as shown in figure 6.10. The fuel stored in the tank at atmospheric pressure is pumped by an electrical feed pump at ca. 0.5 MPa and filtered (0.5 μm) before being pressurized by the high-pressure pump. The fuel is delivered to a rail at a pressure of up to 80 MPa. Accumulators are implemented on both the inlet and outlet sides of the pump (except for the experiments performed with 3-piston pumps) in order to reduce the pulsations due to the pump's single-piston design and to avoid cavitation. The regulation of the rail pressure is automated via a throttle valve so that variations in any other operating parameters do not require manual correction. The fuel pumped through the rail is water-cooled in a heat exchanger before returning to the tank. In the case of the single-piston prototype pump, an auxiliary circuit was necessary to guarantee sufficient lubrication of the cam/pusher system (dashed lines in figure 6.10).

The rail pressure and the pump driving torque (3-piston pumps only) are measured on a shaft-angle-resolved basis in the test rig. In addition to the highly time-resolved measurements, the test rig supplies some other measurements, which are listed below:

- pressure in the low pressure circuit

- low-pressure circuit temperature

- high-pressure circuit temperature

- fuel temperature in the tank

- output mass flow

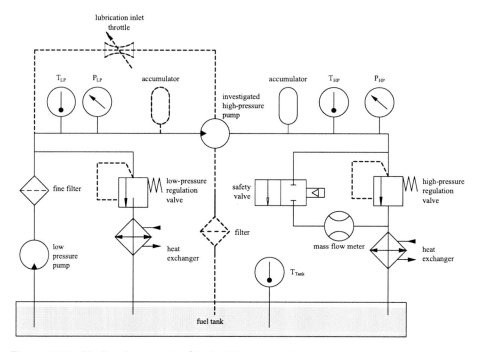

Figure 6.10: Hydraulic test rig (dashed line: single-piston prototype pump only)

Table 6.6 provides details about the sensors used in the test bench. Some of these sensors were specific to the investigated pump (single-piston pump or 3-piston pump).

6.5 Test procedure

The pumps have been tested at operating points corresponding to real operation in an injection system. As the high-pressure pump is supposed to be driven by the camshaft of a GDI-engine, the lowest pump rotation speed has been set to 300 rpm (corresponding to an engine idle speed of 600 rpm). The maximum speed has been set to 2,900 rpm, corresponding to the nominal speed of the asynchronous motor driving the pump in the test bench. For all the experiments, the fuel temperature at the pump inlet was regulated at 20 °C. The following subsection provide more details about the test procedure for each investigated pump.

6.5.1 Single-piston prototype pump

The single-piston prototype pump has been tested in a sequence of operating points shown in figure 6.11. The cycle begins with a system pressure of 20 MPa for a duration of 1 hour. During this hour, the rotation speed is increased stepwise from 300 rpm to 1,000 rpm, 1,600 rpm and 2,900 rpm, before being decreased stepwise

Sensor	Measuring principle	Measuring range	Manufacturer / Type
high-pressure pipe pressure[1]	strain gauge	0..100 MPa	Baumer / PDRD E001.S14B510
high-pressure pipe pressure[2]	strain gauge	0..180 MPa	Bosch B261-260-417-05
low-pressure pipe pressure	strain gauge	0..1 MPa	Sensit / M6015-10A-01/10V
fuel temperature (pipes, tank)	resistance	-50..+400 °C	Electronic Sensor / Pt100A 30/10
fuel mass flow	Coriolis balance	0.1..5 kg/min	Rheonik / RHE 03 GNT
pump driving torque/ driveshaft angle[2]	strain gauge / optical	±20 N · m / 1°	Kistler 4502 A20

[1] single-piston pump only   [2] 3-piston pump only

Table 6.6: Sensors used in the single-piston pump

in the reverse order. Next, the system pressure is set to 30 MPa for 1 hour and to 50 MPa for 2 hours with the same sequential variations in rotation speed. In the investigated rotation speed range (300 rpm to 2,900 rpm), the mean sliding speed in the sliding systems varies between $0.060\,\frac{m}{s}$ and $0.580\,\frac{m}{s}$. The maximum sliding speed varies between $0.094\,\frac{m}{s}$ and $0.911\,\frac{m}{s}$. One cycle lasts approximately 4 hours for a total amount of 272,000 strokes and is repeated as necessary for the desired investigation.

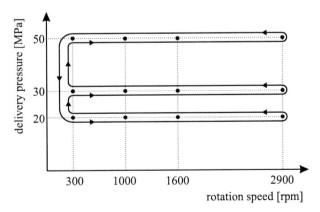

Figure 6.11: Sequence of operating points used for the single-piston pump

6.5.2 3-piston prototype pump

The 3-piston prototype has been tested in a sequence of operating points represented in figure 6.12. This test cycle focus on low rotation speeds, as a low speed leads to poor lubrication in the tribotechnical systems. During a test cycle, the delivery pressure is varied from 20 MPa to 80 MPa and the pump rotation speed from 300 rpm (idle camshaft speed) to 1,300 rpm (middle range camshaft speed). The test cycle lasts approximately 4 hours (220,000 revolutions) and is repeated as necessary for the desired investigation.

The smaller eccentricity of the pump driveshaft leads to a lower rotation speed in comparison to the single-piston pump. In the investigated speed range, the mean sliding speed varies from 0.040 $\frac{m}{s}$ to 0.173 $\frac{m}{s}$. The maximum sliding speed varies between 0.063 $\frac{m}{s}$ and 0.272 $\frac{m}{s}$.

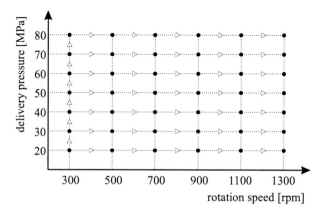

Figure 6.12: Sequence of operating points used for the 3-piston prototype pump

6.5.3 Commercially available pumps

The commercially available pumps have been tested under the same conditions as the 3-piston prototype pump. However, the Bosch HDP1 could only reach 30 MPa without damages (the maximum delivery pressure is 20 MPa according to the manufacturer) and could be tested over four hours only (320,000 revolutions). The Bosch CP1 has been investigated during eight hours (400,000 revolutions).

The relative speed in the sliding systems of the production pumps is higher than in the prototype pumps. In the investigated pump rotation speed range, the mean sliding speed in the tribotechnical systems of the HDP1 varies between 0.066 $\frac{m}{s}$ and 0.286 $\frac{m}{s}$. The maximum sliding speed is between 0.104 $\frac{m}{s}$ and 0.450 $\frac{m}{s}$. The CP1 shows even higher speeds with a mean sliding speed of 0.070 $\frac{m}{s}$ to 0.303 $\frac{m}{s}$ and a maximum sliding speed of 0.110 $\frac{m}{s}$ to 0.476 $\frac{m}{s}$.

7 Results

The results presented in this chapter focus on the friction losses and wear in various prototype and commercially available pumps. The measurements performed with the single-piston prototype pump provide the friction coefficients in the cam/pusher and piston/cylinder systems. The results obtained with the 3-piston prototype pump enable to evaluate the global efficiency of a high-pressure pump with ceramic components in its sliding systems. In addition, a spectrum analysis of the torque measurements helps to characterize the noise emission of the pump with various material pairs in the cam/pusher system. Investigations with production high-pressure pumps provide a reference in order to better evaluate the performance of the prototype pumps developed within the work presented in this document.

7.1 Performance of the reciprocating sliding systems in the single-piston prototype pump

7.1.1 Comparison of the theoretical and measured parameters

In a first step, the measured data has been compared to simulated traces calculated with the equations given in section 3.2. The measurements performed on the test bench lead to results which differ from theoretical signals for several reasons provided in this subsection. The differences are visible in the pressure trace as well as in the forces measured in the sliding systems.

Cylinder pressure
Figure 7.1 shows a real measurement performed at the test bench and its corresponding simulated trace calculated via equations 3.8 and 3.17. The angle 0 °CA corresponds to the bottom dead centre of the piston. An important difference can be seen between the theoretical and the measured data during the compression stroke: the maximum pressure is reached later in the real case. This phenomenon can be explained by the leakage between the piston and cylinder causing a reduction of the mass being compressed in the working chamber. In addition, the elasticity of the whole pump contributes to increase the dead volume, thus extending the time required for compression. The real re-expansion starts before top dead centre because of the leakage in the throttle gap seal: as the piston speed tends to zero near TDC, the instant pump mass flow is lower than the leakage flow between the piston and the cylinder. As a result, the mass to volume ratio in the working chamber decreases, thus causing the expansion of the fuel before

TDC. The elastic deformation of the pump components contributes to extend the re-expansion in comparison to the theoretical data since the dead volume is increased. Consequently, the pump mass flow is significantly reduced.

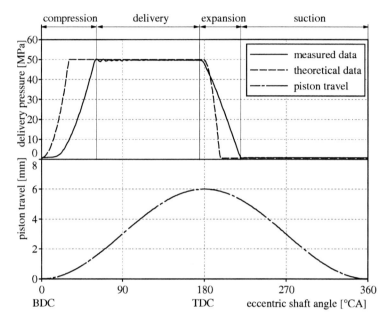

Figure 7.1: Example of cylinder pressure measurement at 50 MPa delivery pressure and 300 rpm pump rotation speed

Forces and friction coefficient in the cam/pusher system

According to the theoretical pressure trace shown in figure 7.1 and assuming that the friction coefficient in the cam/pusher system is constant, it is possible to calculate the friction force in this system (see subsection 6.1.1). The theoretical friction force trace and the data measured at the test bench are shown in figure 7.2. For this example, the delivery pressure has been set to 50 MPa and the theoretical friction coefficient between the cam and the pusher has been set to 0.1. Like the pressure trace, the friction force trace differs from the theoretical data. The differences can be explained by the longer compression and re-expansion strokes mentioned previously (since the contact force is proportional to the cylinder pressure). The change in direction of the real friction force trace (as induced by the pump kinematics, see figure 6.1) occurs later than the theoretical case because of the system elasticity: the change of contact points between the piston and the cylinder only starts after the elastic recovery of the stressed components. This difference is more important after 90 °CA than after 270 °CA because of the higher stress levels before direction change (maximum pressure force reached before 90 °CA, low pressure force around 270 °CA). Another major explanation to the differences that can be observed between the real and theoretical results is that the real friction coefficient in the sliding system is not constant.

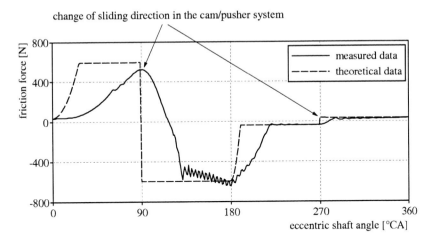

Figure 7.2: Example of friction force measurement in the cam/pusher system at 50 MPa delivery pressure and 300 rpm rotation speed (theoretical data calculated with a friction coefficient of 0.1)

Figure 7.3 gives an example of instant friction coefficient between the cam and the pusher. This coefficient is calculated with the measured friction and contact forces via equation 4.8. The figure enables to assert that the friction coefficient varies significantly over a single driveshaft revolution (in this example: between 0 and 0.13). Its value is affected by several parameters such as the sliding speed and contact pressure, as the following results in this chapter will show. The friction coefficient is equal to zero after 90 °CA and 270 °CA because of the sliding direction change in the cam/pusher system (friction force equal to zero, see figure 7.2).

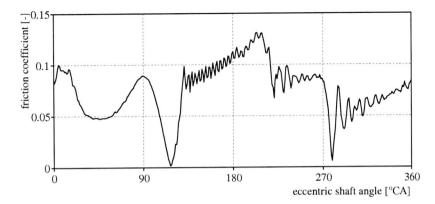

Figure 7.3: Calculated instant friction coefficient in the cam/pusher system based on the data plotted in figures 7.1 and 7.2

In order to compare the performance of the investigated material pairs (especially regarding friction losses), it is necessary to calculate representative values for each measurement. Time-averaging the value of the friction coefficient is not satisfactory as the weighting of the low friction force range (between ca. 200 °CA and 360 °CA) would be the same as the weighting of the high friction force range (between ca. 0 °CA and 200 °CA). For that reason, the value of the friction coefficient as provided in this document is calculated as a function of the friction force $F_{friction}$, the contact force $F_{contact}$ and the relative motion $ds_{friction}$ via following equation:

$$\mu = \frac{\int F_{friction} \cdot ds_{friction}}{\int F_{contact} \cdot ds_{friction}} \tag{7.1}$$

This integral friction coefficient better describes the friction losses with the investigated material pairs since it is linked to the work of friction as defined in the subsection 4.3.2.

Forces and friction coefficient in the piston/cylinder system

Figure 7.4 provides an example of friction force measurement between the piston and the cylinder as a function of the eccentric shaft angle. The phenomenon of piston rocking can be identified in the figure: the friction force decreases slightly after 90 °CA. The change of sliding direction between the piston and the cylinder occurs shortly after 180 °CA. The elasticity of the pump explains the delayed inversion of the friction force. After TDC, the pressure valve closes but the fuel maintains pressure on the piston during the process of re-expansion until ca. 220 °CA, thus explaining the high friction force level visible in the figure between 180 °CA and 220 °CA.

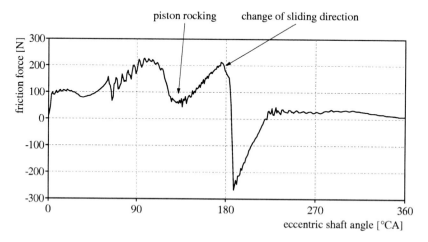

Figure 7.4: Example of friction force measurement in the piston/cylinder system at 50 MPa delivery pressure and 300 rpm rotation speed

As the piston changes its contact points with the cylinder after 90 °CA and

270 °CA, the contact force between both parts decreases significantly. This leads to high friction coefficient peaks as exemplary shown in figure 7.5. These peaks are not representative of the friction losses between the piston and the cylinder. Consequently, the friction coefficient in this tribotechnical system has been calculated via the same method as for the cam/pusher system (see equation 7.1).

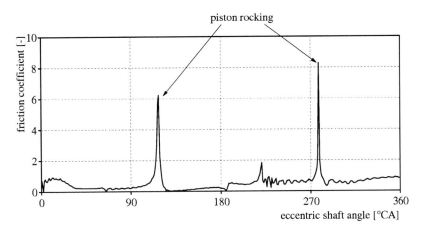

Figure 7.5: Calculated instant friction coefficient in the piston/cylinder system based on the data plotted in figures 7.2 and 7.4

7.1.2 Friction in the investigated sliding systems

This subsection shows the friction coefficients measured in the cam/pusher system and the friction work in the piston/cylinder system. Unless mentioned otherwise, each material pair has been tested during 16 hours, corresponding to 4 cycles as described in figure 6.11. The results provided by the following figures have been acquired during the last measuring cycle and averaged over 50 revolutions of the pump driveshaft. Apart from some specific comparisons, all the results shown in this subsection have been obtained while using isooctane as delivered fuel and lubricant.

Cam/pusher system
Several material combinations have been tested in the cam/pusher system. Since changing the material pair in one of the two investigated sliding systems may affect the performance of the other, only self-mated silicon carbide has been used in the piston/cylinder system.

In a first step, investigations have been performed in order to compare the friction losses of self-mated silicon carbide with different surface roughnesses. Figure 7.6 presents the friction coefficients as a function of the delivered pressure and the pump rotation speed. The friction coefficients tend to decrease with increasing rotation speed as the increasing sliding speed in the sliding systems improves the lubrication (see Stribeck curve in figure 4.9). Increasing the delivery pressure from 20 MPa to 50 MPa does not affect the results in a significant way.

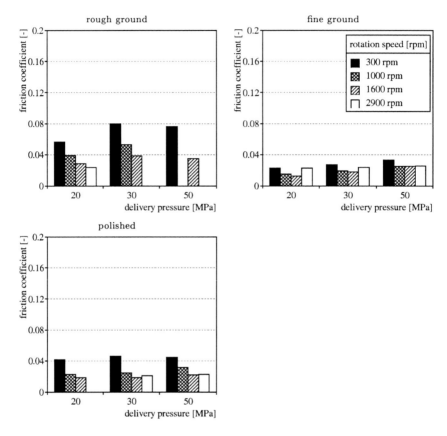

Figure 7.6: Comparison of silicon carbide material pairs with various surface roughnesses in the cam/pusher system (fuel: isooctane)

The results obtained with the rough ground surfaces show higher friction coefficients when compared to the other material pairs. In addition, not all the operating points could be investigated at 30 MPa 50 MPa delivery pressure because of the high and irregular friction forces observed during the test. This can be explained by the abrasive effect of the particles released during the run-in, which is considerably longer in comparison to the other tests performed.

Despite its very smooth surface (R_a=0.006 µm), the polished material pair does not suggest an improved performance in comparison to the fine ground material pair. This observation can be explained by the fact that the surfaces are already smoothed at the beginning of the test, so that the fit between the parts cannot be improved by initial wear. Due to unstable operation, no measurement has been performed at 20 MPa and 2,900 rpm.

The fine ground material pair shows the best global performance with a friction coefficient in the range of 0.015 to 0.045 depending on the delivery pressure and rotation speed considered. In contrast to the polished pair the surface of the fine ground pair shows peaks and valleys at microscopic scale. During the run-in, these asperities are smoothened in order to get a better stress distribution on the surface, thus explaining the improved performance of the fine ground pair in comparison to the polished pair. Unless mentioned otherwise, all the following results presented in this document have been obtained with fine ground material pairs.

Further experiments were carried out in order to evaluate the performance of self-mated sialon and silicon carbide combined with AISI 52100. Figure 7.7 shows the friction coefficients of the two material pairs as a function of the pump delivery pressure and rotation speed. At 20 MPa and 30 MPa, the sialon pair shows results which are close to the fine ground silicon carbide pair with a friction coefficient in the range of approximately 0.015 to 0.04. At 50 MPa delivery pressure, the pump rotation speed has a greater influence on the friction losses: at 300 rpm, the friction coefficient is higher than with self-mated silicon carbide, whereas at higher rotation speed, the friction coefficient is slightly lower (between 0.01 and 0.02). The range of the friction coefficient observed with the self-mated material pairs suggests a state of mixed friction between the components of the sliding system (see table 4.1).

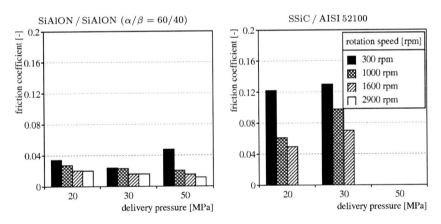

Figure 7.7: Friction coefficient in the cam/pusher system with self-mated sialon or silicon carbide combined to AISI 52100 (fuel: isooctane)

The investigation of the silicon carbide combined with AISI 52100 was unsuccessful since the friction coefficient tended to increase over time at a single operating point. For this reason, no measurements were performed at 50 MPa delivery pressure or at high rotation speeds. The test was aborted after 3 hours of operation instead of 16 hours for a full test. In contrast to the other material pairs, no stabilization of the friction forces was observed in this time range. At 20 MPa and 30 MPa, the friction coefficient is in the range of 0.05 to 0.13, thus suggesting a boundary friction state between the cam and the pusher.

As explained in subsection 6.2.4, micro-texturing the surface of one sliding part should enhance the lubrication of the contact by providing a storage volume for the particles released by wear. Figure 7.8 shows the results obtained with textured silicon carbide cams against silicon carbide or AISI 52100. An improved performance can be observed for the textured self-mated silicon carbide pair in comparison to the reference test shown in figure 7.6 (self-mated silicon carbide, fine ground). The friction coefficient is approximately 0.01 at all the investigated operating points and does not seem to be influenced by the pump rotation speed or the delivery pressure.

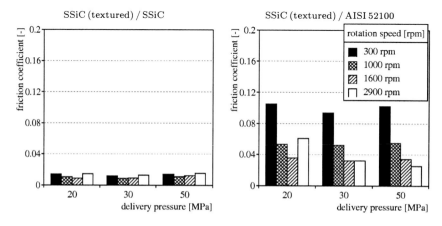

Figure 7.8: Friction coefficient in the cam/pusher system with various material pairs with a texture on the cam (fuel: isooctane)

The performance of silicon carbide in combination with AISI 52100 is also considerably improved by texturing the cam. In contrast to the test without texture, measurements at all operating points were possible without encountering any increase of the friction forces over time. The pump rotation speed strongly influences the friction coefficient level. At 300 rpm, the friction coefficient is approximately 0.1, and decreases to values between 0.02 and 0.06 at higher rotation speeds. The results do not suggest an influence of the pump delivery pressure on the friction coefficient in the cam/pusher system.

The results provided by the previous figures show friction coefficients which are more or less influenced by the pump rotation speed or by the pump delivery pressure. These parameters describe the operating point of the pump but do not correspond to the stress applied to the tribotechnical systems. In the cam/pusher system, the sliding speed is not constant over one driveshaft revolution as it depends on the angular position of the eccentric shaft (see equation 3.5). The surface pressure $p_{surface}$ in the contact also varies: its value can be calculated by summing the friction force between the piston and the cylinder F_z, the force applied to the piston by the cylinder pressure $p_{cylinder}$ and the retaining spring force F_s (A_{piston}

and A_{pusher} designate the area of the piston and the pusher, respectively):

$$p_{surface} = \frac{(F_z + p_{cylinder} \cdot A_{piston} + F_s)}{A_{pusher}} \tag{7.2}$$

The combination of the instant sliding speed, surface pressure and friction coefficient (figure 7.3) enable to generate a scatter plot as exemplary shown in figure 7.9. By merging all the measurements performed with a single material pair (20 MPa to 50 MPa and 300 rpm to 2,900 rpm), the scatter plot covers the range of approximately $0\frac{m}{s}$ to $0.9\frac{m}{s}$ (instant sliding speed) and 0 MPa to 12 MPa (instant surface pressure). The plot created with the instant values show a decrease of the friction coefficient with increasing sliding speed and increasing surface pressure. These observations can be explained by following reasons: a higher relative speed between the bodies enhances lubrication as shown in figure 4.9. In addition, a better stress distribution or tribochemical reactions with increasing surface pressure may reduce the friction losses in the system.

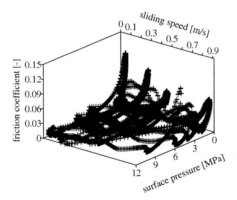

Figure 7.9: Instant friction coefficient in the cam/pusher system as a function of the surface pressure and the relative sliding speed (measured data)

Taking the operating parameters and the friction coefficient range into account, it can be inferred that the friction states observed correspond to boundary and mixed friction. The fluid friction can be neglected since no significant increase of the friction coefficient at high rotation speeds could be observed. For better readability, the scatter plots have been approximated by surfaces with the following equation form:

$$f(x, y) = a + b \cdot e^{-c \cdot x} \cdot e^{-d \cdot y} \tag{7.3}$$

In this equation, x and y designate the sliding speed and the surface pressure between the cam and the pusher. a, b, c and d designate constants, which have been determined via the surface fitting tool provided by the software Matlab.

Figure 7.10 shows the instant friction coefficient measured with self-mated silicon carbide (with and without texture on the cam), sialon and silicon carbide combined with AISI 52100. It can be inferred from these diagrams that the averaged friction coefficients presented in figures 7.6, 7.7 and 7.8 lead to results which are comparable to the instant measurements.

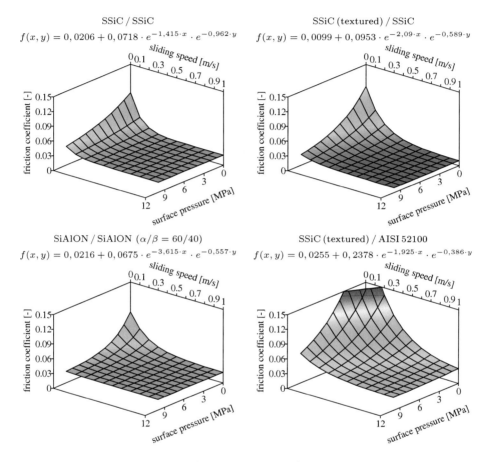

Figure 7.10: Instant friction coefficient in the cam/pusher system with various material pairs: surface fitting (fuel: isooctane)

Figure 7.10 enables to notice that texturing the cam of the self-mated silicon carbide pair leads to a slightly increase of the friction coefficient at slow speed and low surface pressure. However, as the friction coefficient tends to a lower value with increasing speed and pressure, the averaged performance is better in comparison to the untextured material pair, as shown in figure 7.8. The self-mated sialon pair shows very similar results to those obtained with silicon carbide. In contrast, the combination of a textured silicon carbide cam with a pusher made of AISI 52100 shows a higher friction coefficient over the entire investigated range, as could be inferred from the results presented in figure 7.8. The influence of an increasing relative sliding speed and surface pressure on the friction coefficient is significantly higher for this material pair.

In order to validate the results obtained by using isooctane as a reference fuel, complementary measurements have been performed with commercial gasoline (RON 95) and different gasoline/ethanol mixtures. Figure 7.11 shows the results obtained with gasoline lubricated self-mated silicon carbide and textured silicon carbide combined to AISI 52100 in the cam/pusher system. The results obtained with self-mated silicon carbide only show a slightly decrease of the friction coefficient in the cam/pusher system at mid-range rotation speed and 50 MPa delivery pressure. In the case of silicon carbide (textured) combined to AISI 52100, a significant improvement of the performance can be observed when compared to the results displayed in figure 7.8. The delivery pressure shows a strong influence on the friction coefficient. At 50 MPa, the results are similar to self-mated silicon carbide at low and mid-range rotation speeds. The improved performance can be explained by the higher viscosity of gasoline in comparison to isooctane. The content of various chemical components which improve the lubricating properties of gasoline may also contribute to the decreasing friction coefficient with increasing delivery pressure.

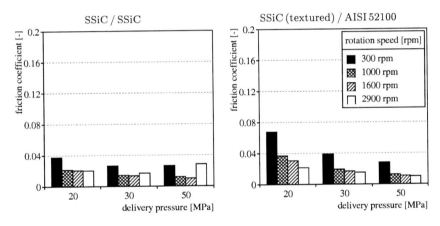

Figure 7.11: Friction coefficient in the cam/pusher system with various material pairs (fuel: gasoline)

Figure 7.12 shows the friction coefficient obtained at 50 MPa delivery pressure with various fuels. The material pair used in the cam/pusher system for this test is self-mated silicon carbide. The use of ethanol or ethanol/gasoline mixtures as delivered fuel and lubricant leads to a decreasing friction coefficient. With 25 % ethanol or more, the friction coefficient in the cam/pusher system remains below 0.02 at all rotation speeds. The differences observed with the various fuels used for lubrication can be essentially explained by their different viscosity (0.348 mPa·s for isooctane, 0.65 mPa·s for gasoline and 1.2 mPa·s for ethanol). However, since a significant reduction of the friction coefficients can be observed with a very low ethanol ratio in gasoline (E25 with 25 % ethanol) and since the investigated gasoline also contains 5 % ethanol, tribochemical reactions between the surfaces and the fuel should not be excluded.

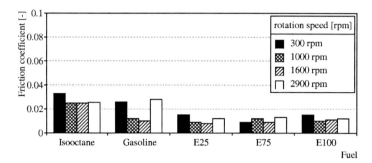

Figure 7.12: Friction coefficient in the cam/pusher system with various fuels at 50 MPa delivery pressure (cam/pusher material: SSiC / SSiC)

The results presented in figures 7.6 to 7.12 (except the results corresponding to untextured silicon carbide combined with AISI 52100) were obtained during the last testing cycle. At the beginning of the test, the investigated material pairs go through a run-in phase with high friction coefficients, before reaching stable performance. This effect can be explained by various friction mechanisms and by the release of wear particles in the sliding system. The particles have an abrasive action on the surfaces and cause high friction forces and wear. Figure 7.13 shows the evolution of the friction coefficients for various material pairs as a function of the sliding distance. These plots correspond to trend lines of all the friction coefficients measured during the stress collective (all operating points are included). All the results presented in this figure were obtained with isooctane as delivered fuel. The curve corresponding to the combination of textured silicon carbide with AISI 52100 stops after a sliding distance of 6,500 m because gasoline was used for complementary investigations after this point.

The friction coefficients of all material pairs decrease significantly at the beginning of the tests except in the case of untextured silicon carbide in combination with AISI 52100. As previously mentioned, the tests with this material combination were unsuccessful because of the increasing friction forces over time at a single operating point. The SEM analysis, which is presented later in this section, provides an explanation to this observation. The self-mated silicon carbide and sialon pairs show a similar run-in behaviour with a strong decrease of the friction coefficient during the first 5,000 m sliding distance and a slower decrease after 5,000 m. The textured material pairs show a faster run-in since the most significant reduction of their friction coefficients is already achieved after approximately 2,000 m. This observation validates one of the expected effects of the texture: the micro-structure stores the particles released by wear, thus limiting their abrasive action during the run-in phase.

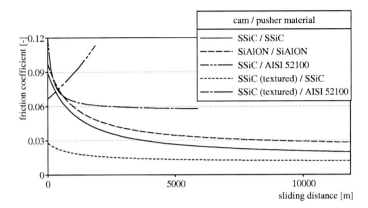

Figure 7.13: Run-in in the cam/pusher system with various material pairs (fuel: isooctane)

Piston/cylinder system

Two material pairs have been tested in the piston/cylinder system: self-mated silicon carbide and the combination of a piston made of AISI 52100 and a cylinder made of silicon carbide. Self-mated silicon carbide has been used in the cam/pusher system for both tests. The measurements performed at high rotation speeds show high oscillations of the friction force between the piston and the cylinder (see figure 7.14). These oscillations are caused by vibrations of the pump. Their amplitude is significantly higher than the actual friction force and filtering the signal leads to important inaccuracies. Therefore, only the results of measurements performed at 300 rpm are displayed in this document. Comparing the two investigated material pairs in the piston/cylinder system via the friction coefficient may lead to inaccuracies as the contact forces between the piston and the cylinder are a function of the friction force in the cam/pusher system. The friction work better describes the performance of the material pairs as it is a function of the friction force between the piston and the cylinder only (see subsection 4.3.2).

Figure 7.15 depicts the friction work in the piston/cylinder system at 300 rpm as a function of the pump delivery pressure. Both material pairs show similar performance with a friction work in the range of 0.2 J to 0.35 J between 20 MPa and 50 MPa delivery pressure. These values are very low when compared to the pressure-volume work, which is approximately 12 J, 18 J and 28 J at 20 MPa, 30 MPa and 50 MPa delivery pressure, respectively: the friction work in the piston/cylinder system represents approximately 1.5 % of the effective pump work (for both materials). As a result, both material pairs offer very good results regarding friction losses for the use in the piston/cylinder system.

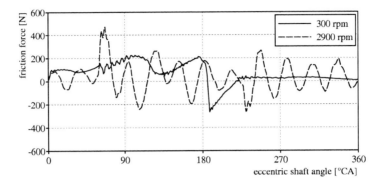

Figure 7.14: Friction force measured in the piston/cylinder system at 300 rpm and 2900 rpm (50 MPa delivery pressure; fuel: isooctane)

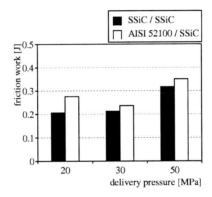

Figure 7.15: Friction work over one eccentric shaft revolution in the piston/cylinder system at 300 rpm (fuel: isooctane)

7.1.3 Surface analysis

Every cam and pusher was analyzed with a scanning electron microscope (SEM) in order to evaluate wear and to better understand the results presented in the previous subsection. Figure 7.16 shows the evolution of the surface of a fine ground silicon carbide cam before the test, after 13,000 m (16 hours of operation) and after 65,000 m (80 hours of operation) sliding distance. A clear difference can be observed between the pictures of the cam before the test and after 13,000 m: the surface is smoothened during the run-in. It can be supposed that the presence of cavities on the surface improves the lubrication of the sliding system since they keep fuel in the contact. The smoothing of the surface asperities supports a better stress distribution (better fit between the parts) and thus improves the performance of the material pair, as mentioned in the previous subsection. There is no visible change for longer test durations (65,000 m) and the specific wear rate could not be calculated because of the very low wear volume.

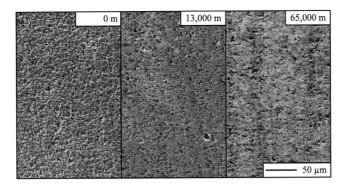

Figure 7.16: SEM pictures of silicon carbide cams used in combination with silicon carbide pushers (fine ground) after increasing test durations (0 m to 65,000 m sliding distance)

The evolution of the polished and rough ground silicon carbide cams is shown in figure 7.17. No visible change of the surface state can be observed in the case of the polished pair. Despite its smoother surface, the polished material pair did not perform better than the fine ground pair. This can be explained by the very smooth surface before the test: the fit between the parts can barely be improved by initial wear. The rough ground pair shows high asperities even after 13,000 m sliding distance. Even if a rough surface could support a better part fitting, the increased release of wear particles during the run-in may explain the high and irregular friction forces observed during the tests. As a result, the fine ground pair provides a good compromise between fast run-in, good fit between the surfaces and stable performances.

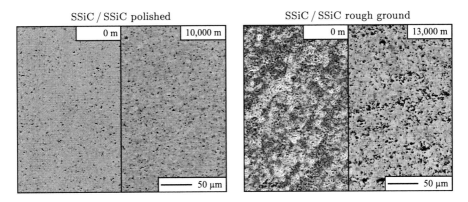

Figure 7.17: SEM pictures of the polished and the rough ground silicon carbide cams before and after test

Shown in figure 7.18 are SEM pictures of the sialon cam surface before and after testing. On the left, horizontally-oriented grooves left by the grinding process can clearly be identified. The experimentally measured maximum depth of these grooves is approximately $0.5\,\mu m$. On the right, a smoothing of the surface in the sliding direction (vertical) can be seen. However, some horizontal grooves remain in the material, which means that the wear in the sliding system is less than $0.5\,\mu m$ after $13{,}000\,m$ sliding distance. This allows to estimate the specific wear rate as defined by equation 4.10. The worn surface area is $940\,mm^2$, thus leading to a maximum wear volume of $0.47\,mm^3$. The average contact force in the system during the tests is approximately $3{,}000\,N$. As a result, the specific wear rate is lower than $1.6{\cdot}10^{-8}\,\frac{mm^3}{N{\cdot}m}$, well below the examples provided in figure 4.8.

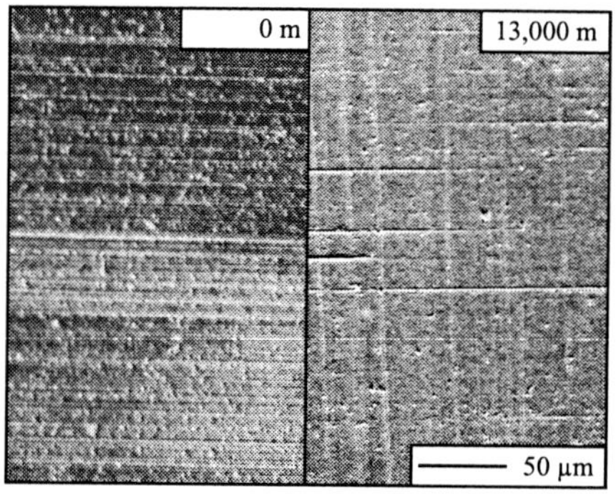

Figure 7.18: SEM pictures of the sialon cam before and after investigation

Figure 7.19 explains what has been observed during the run-in of the silicon carbide cam combined with the AISI 52100 pusher. A steel oxide layer can be seen on the surface of the untextured cam (white arrows in the picture on the left) after only $2{,}000\,m$ sliding distance (between $20\,MPa$ and $30\,MPa$ delivery pressure). This layer indicates an adhesion mechanism occurring in the sliding system and explains the increasing friction coefficient over time shown in figure 7.13. Investigations performed with a tribometer in [98] have led to similar observations. No steel layer has been observed on the surface of the textured cam (figure 7.19 in the middle). This confirms that the texture enhances lubrication and prevents the adhesion mechanism. The surface of the textured silicon carbide cam is smoothened and the grains of the ceramic can be identified in the picture. As silicon carbide is 3.2 times harder than AISI 52100 ($2540\,HV$ for the investigated silicon carbide, $790\,HV$ for the hardened AISI 52100), such an observation can only be explained by chemical reactions on the surface. This phenomenon was also observed in [98]. Tribochemical reactions may also be an explanation to the positive influence of a high surface pressure on the friction coefficient (see figure 7.10). The grooves visible in the picture of the AISI 52100 pusher (on the right) are left by wear particles. However no severe wear can be observed on the pusher.

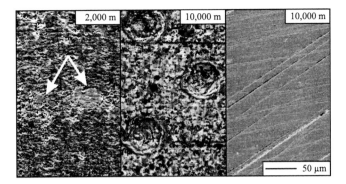

Figure 7.19: SEM pictures of silicon carbide cams tested with pushers made of AISI 52100 (left: cam without texture; middle: cam with texture; right: pusher used against the textured cam)

7.2 Application in a 3-piston pump

This section presents the results obtained with the 3-piston prototype pump with various material pairs in the cam/pusher system. The only material combination used in the piston/cylinder system is AISI 52100 (piston) combined with silicon carbide (cylinder). This material pair provides very good performance as shown in the previous section. Moreover, using pistons made of steel offers an easier manufacturing in comparison to pistons made of silicon carbide. The pump has been operated with gasoline and complementary investigations have been performed with ethanol.

7.2.1 Comparison of theoretical and measured parameters

Like in the case of the single-piston prototype pump, differences can be observed between the theoretical and measured signals with the 3-piston prototype pump. The main highly time-resolved measurements leading to the results shown in this section are the cylinder pressure and the driving torque.

Cylinder pressure

Figure 7.20 shows an example of pressure trace measured at 300 rpm rotation speed and 80 MPa delivery pressure. Only the signal of one cylinder is displayed in order to facilitate the reading. The differences between the theoretical and measured data are similar to those observed with the single-piston pump. The compression is delayed by the leakage in the throttle gap seal and by the elastic deformation of the various parts in the pump. The re-expansion begins before TDC and is extended, thus causing volumetric losses. The shape of the pressure traces measured in each cylinder slightly differs from each other because of the dimensional tolerance on the parts. This affects the dead volume size of each work chamber and the leakage mass flow in the throttle seal gaps.

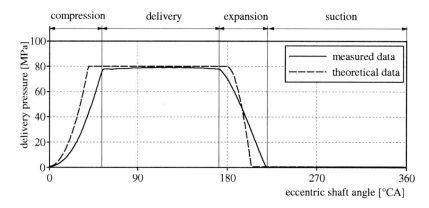

Figure 7.20: Example of cylinder pressure measurement in the 3-piston prototype pump at 80 MPa delivery pressure and 300 rpm rotation speed

Driving torque

In contrast to the single-piston prototype pump, the 3-piston pump is not fitted with force sensors in its sliding systems. As a result, it is not possible to measure the friction coefficients between the cams and the pushers or between the pistons and the cylinders. The friction losses can be estimated via the pump mechanical efficiency. The driving torque is measured against the eccentric shaft angle for that purpose. Figure 7.21 gives an example of torque measurement at 300 rpm rotation speed and 80 MPa delivery pressure. In the ideal case (identical pressure trace in each cylinder, ideal geometry of all the parts in the pump), the torque trace should repeat every 120 °CA due to the 3-piston design. However, the pressure traces are not identical, thus affecting the shape of the driving torque. Consequently, the theoretical data provided in figure 7.21 has been calculated as follows: the pressure traces enable to calculate the force applied by each pusher on the eccentric lobe. Assuming that there is no friction in the sliding systems, the resulting force on the eccentric lobe enables to determine the driving torque required to generate the measured pressure traces (see balance of forces in subsection 6.1.2). The difference between the theoretical and the measured torque (hatched area in figure 7.21) highlights the friction losses in the sliding systems of the pump.

7.2.2 Mechanical efficiency with various material pairs in the cam/pusher system

The energy losses between the pump inlet and outlet can be explained by several factors as mentioned in subsection 3.2.3. In order to avoid the influence of fluctuating volumetric and hydraulic losses between each experiment, the performance of the investigated material pairs has been analyzed by considering the pump mechanical efficiency only. The mechanical efficiency can be calculated via the ratio between the indicated power (see equation 3.9) and the input power (equation 3.11). Unless mentioned otherwise, the signals used for the calculation have

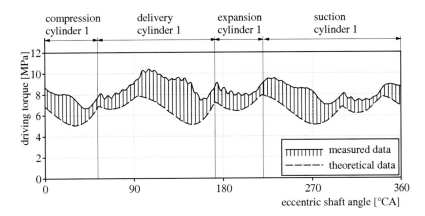

Figure 7.21: Example of torque measurement of the 3-piston prototype pump at 80 MPa delivery pressure and 300 rpm rotation speed

been measured after 5,000 m sliding distance in the sliding systems and have been averaged over 50 cycles. Some of the diagrams do not cover the entire speed and pressure range. This is generally due to unstable operation preventing reliable measurements at specific operating points.

In this subsection, the figures provide the mechanical efficiency of various material pairs in the cam/pusher system. The delivery pressure range is 20 MPa to 80 MPa and the speed range is 300 rpm to 1,300 rpm. The eccentricity of the driveshaft in the 3-piston pump is 2 mm instead of 3 mm in the single-piston pump. As a result, the investigated speed range with the 3-piston pump corresponds to a rotation speed range of 200 rpm to 867 rpm with the single-piston pump. Furthermore, since the contact surface between the cams and the pushers is smaller in comparison to the single-piston prototype pump, the delivery pressure range of 20 MPa to 80 MPa corresponds to the range of 10 MPa to 64 MPa with the single-piston pump. The lower relative speed range and the higher specific surface pressure of the 3-piston prototype pump lead to more critical testing conditions than with the single-piston pump.

Figure 7.22 shows the mechanical efficiency obtained with silicon carbide based material pairs in the cam/pusher system. The delivered fuel and lubricant is gasoline. The diagrams show an improved performance with increasing rotation speed. This observation confirms the results obtained with the single-piston pump: a higher pump rotation speed supports a better lubrication since the relative sliding speed between the bodies of the tribotechnical systems increases. It can be noticed from the figure that a higher delivery pressure also leads to a better mechanical efficiency. This can be explained by several factors. A share of the pump mechanical losses are barely dependent on the delivery pressure (friction losses in the slide ring seals for example). As the pressure induced forces increase, the share of the constant friction losses decreases. Consequently, the pump mechanical efficiency increases with higher delivery pressures (assuming that the friction coefficients in the sliding systems are constant). Furthermore, the results obtained with the single-piston pump have shown that a higher contact pressure between the cam and the pusher leads to a lower friction coefficient, especially at low sliding speeds.

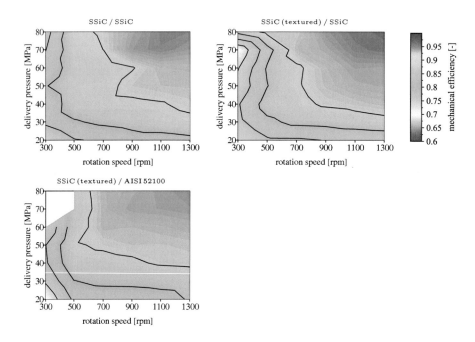

Figure 7.22: Pump mechanical efficiency with silicon carbide in the cam/pusher system (fuel: gasoline)

In the investigated speed and pressure range, the self-mated silicon carbide pair in the cam/pusher system enables to achieve a mechanical efficiency of approximately 0.75 to more than 0.95, depending on the operating point considered. An improvement at rotation speeds above 500 rpm can be noticed with the textured self-mated silicon carbide pair. At lower speed, the untextured pair performs the best. This confirms the results obtained with the single piston-pump in figure 7.10 and can be explained by a better stress distribution on the untextured cam (the surface is reduced by 20 % with the texture). At higher sliding speed, the improved lubrication induced by the micro-structure outweighs the effect of the smaller contact area. The mechanical efficiency measured with the combination of silicon carbide with AISI 52100 is slightly lower than with the self-mated pairs with values between ca. 0.7 and 0.95.

Figure 7.23 shows the mechanical efficiency of the 3-piston pump with various self-mated sialon pairs in the cam/pusher systems. The pump rotation speed affects the results in the same way as the silicon carbide based pairs. However, the delivery pressure has a different influence on the mechanical efficiency when compared to the silicon carbide based pairs: the maximum mechanical efficiency is reached within the investigated pressure range and higher delivery pressures lead to a decreased efficiency. This can be explained by the lower hardness and thermal conductivity of sialon in comparison to silicon carbide.

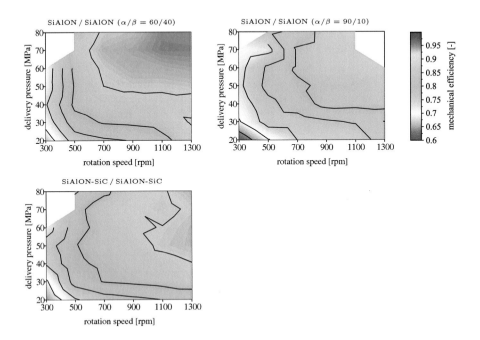

Figure 7.23: Pump mechanical efficiency with sialon in the cam/pusher system (fuel: gasoline)

The sialon pair with the α/β-phase ratio of 60/40 shows the best global performance with a mechanical efficiency between 0.65 and over 0.95. In the investigated speed and pressure range, the optimal operating point is reached at a speed of ca. 1,200 rpm and a delivery pressure of 70 MPa. Despite its higher hardness, the sialon pair with the α/β-phase ratio of 90/10 leads to a lower mechanical efficiency with values between ca. 0.6 and 0.9. Moreover, due to irregular performances and high torque peaks, some operating points could not be investigated (especially at high speed and high delivery pressure). An explanation to this observations is provided in subsection 7.2.5. At delivery pressures below 50 MPa, the SiAlON-SiC composite shows similar performance to sialon with an α/β-phase ratio of 60/40. However, the composite does not suggest any improvement of the mechanical efficiency above 50 MPa delivery pressure.

The results obtained with various silicon nitride pairs in the cam/pusher system are depicted in figure 7.24. Due to high and irregular torque levels, the two self-mated pairs (with and without carbon nanotubes) could not be tested under the same conditions as the other investigated material pairs. The experiment with the self-mated silicon nitride was interrupted after 5,000 m sliding distance since no improvement has been observed over time. The self-mated pair containing carbon nanotubes (CNTs) was tested over 1,700 m sliding distance only. Both self-mated pairs show a mechanical efficiency which is lower than silicon carbide or sialon based material pairs (0.6 to ca. 0.9 depending on the operating point). Using silicon nitride in combination with AISI 52100 does not provide better global performances in comparison to the self-mated pair. The mechanical efficiency is even lower in the mid-range delivery pressure (around 0.8). However, this material pair could be tested under the same conditions as the silicon carbide and sialon based material pairs, thus allowing a direct comparison with the other results shown in this document. The combination of silicon nitride containing CNTs and AISI 52100 leads to very good results since a mechanical efficiency of more than 0.95 is reached at speeds above 900 rpm and delivery pressures higher than 50 MPa. The improved performances can be explained by the following phenomena. First, the deformation capability of the nanotubes enable to better distribute the local stresses on the surface of the bodies. In addition, the nanotubes may improve the lubrication by the formation of CNTs based wear particles acting as a solid lubricant in the contact [32]. Tribochemical reactions in the system cannot be excluded either (see subsection 7.2.5).

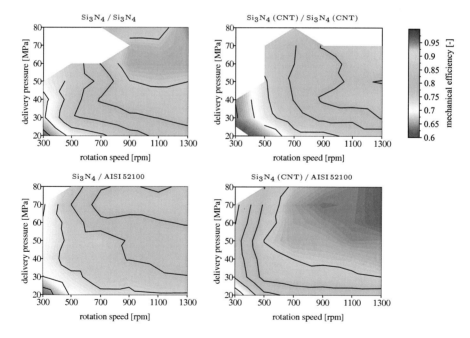

Figure 7.24: Pump mechanical efficiency with silicon nitride in the cam/pusher system (fuel: gasoline)

Complementary measurements have been performed with ethanol as delivered fuel and lubricant. Figure 7.25 provides the pump mechanical efficiency with textured silicon carbide (self-mated and in combination with AISI 52100) and self-mated sialon (α/β-phase ratio of 60/40 and 90/10) in the cam/pusher system. All the investigated material pairs show a very good mechanical efficiency over the entire investigated speed and pressure range with values between 0.8 and more than 0.95. In contrast to measurements performed with gasoline, the pump rotation speed barely affects its mechanical efficiency. This confirms the results obtained with the single-piston pump (see figure 7.12): the higher dynamic viscosity of ethanol in comparison to gasoline and possible tribochemical reactions lead to a significant improvement of the lubrication, even at low rotation speeds.

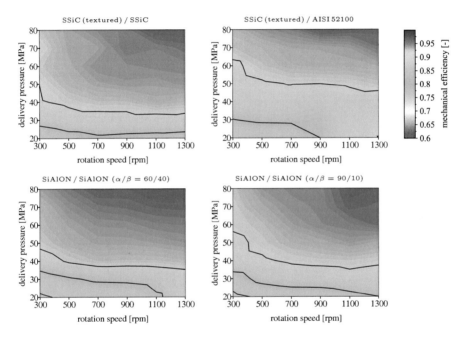

Figure 7.25: Pump mechanical efficiency with ethanol as delivered fuel and lubricant

7.2.3 Cyclic variations and noise emission

The friction in the sliding systems may induce noise emission, especially at low rotation speeds. The experiments were not recorded with a microphone because of the many other noise emitting sources in the test bench (electrical engine, fans, etc). However, an analysis of the torque traces helps to estimate the loudness of the pump operation with the investigated material pairs.

Oscillations of the friction force between two bodies may occur if the friction coefficient decreases with increasing relative speed (as observed with the investigated components, see figure 7.10). This leads to unstable motion at low sliding speeds, and may cause "stick and slip" (alternating idling and sliding phases). The oscillation frequencies generated by friction in the tribotechnical systems are influenced by the properties of the system ifself. However, changing the sliding speed between the bodies may affect the amplitude of the oscillations but does not influence their frequency [67].

The Fourier transform of the torque signal provides the oscillation amplitude of the pump driving torque as a function of the frequency. The frequency range is limited by the sampling rate of the signal. According to the Nyquist-Shannon theorem, the maximum bandwidth resulting from the Fourier transform is equal to the half of the signal sampling rate [18]. Assuming that the pump rotation speed is 500 rpm (corresponding to 8.33 Hz) and that 360 points are measured over one driveshaft revolution, the sampling rate is 3,000 Hz. As a result, the maximum frequency, which can be calculated via Fourier transform at 500 rpm pump rotation speed is 1,500 Hz. The figures in this subsection show the torque spectrum with various material pairs in the cam/pusher system at 500 rpm and 80 MPa (except for the self-mated silicon nitride pairs). The differences between the torque traces may not be caused by the different behaviour of the investigated material pairs only. Fluctuating pressure traces (due to volumetric losses for example) may also affect the oscillations of the torque signal. Therefore, the theoretical torque (calculated with measured pressure traces, see subsection 7.2.1) has also been analyzed via Fourier transform for each investigated material pair. The results have been subtracted to the data obtained with the real torque measurements. Consequently, it can be assumed that the oscillations depicted in this subsection are caused by friction only.

Figure 7.26 shows the torque frequency spectrum over 100 successive cycles obtained with silicon carbide pairs in the cam/pusher systems. The frequency range displayed in the figure is 20 Hz to 400 Hz. Frequencies below 20 Hz are not audible and no significant signal could be identified at higher frequencies. The amplitude scale of the diagrams (0 N·m to 0.4 N·m) corresponds to approximately 0 % to 5 % of the average pump driving torque over one cycle. Despite the high delivery pressure and the low rotation speed, self-mated silicon carbide shows very low friction induced oscillations of the driving torque. Oscillations occur mainly around 125 Hz and 250 Hz with an amplitude lower than 0.25 N·m. Texturing the cam when using self-mated silicon carbide does not affect the results in a significant way since the same oscillation frequencies and amplitudes can be observed. Using a textured silicon carbide cam in combination with AISI 52100 leads to similar results. The cyclic variations are low for all the silicon carbide pairs. Only periodical fluctuations of the oscillation amplitudes around 125 Hz and 250 Hz can be observed. During the experiments, no particular noise emission was noticed.

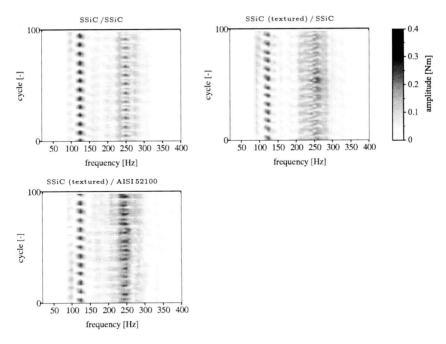

Figure 7.26: Spectrum analysis of the pump driving torque over 100 revolutions with silicon carbide in the cam/pusher system (fuel: gasoline; rotation speed: 500 rpm; delivery pressure: 80 MPa)

The results from the investigations performed with sialon pairs in the cam/pusher systems are provided by figure 7.27. The amplitude scale of the diagrams corresponds to approximately 0 % to 10 % of the average pump driving torque over one cycle. Both pure self-mated sialon pairs (α/β-phase ratio of 60/40 and 90/10) show similar torque frequency spectra. The oscillations around 250 Hz are significantly higher than those observed with the silicon carbide pairs (up to 0.7 N·m). During the tests, noise emission induced by the friction in the sliding systems could be clearly identified. In addition, the pure sialon pairs show cyclic variations which are considerably higher than with silicon carbide pairs. The amplitude at 125 Hz shows periodic fluctuations like the silicon carbide pairs. But at higher frequencies (especially around 250 Hz), high non-periodic fluctuations of the oscillation amplitude and slightly frequency shifts can be identified. These spectra changes were audible during investigation at the test bench. The torque measured with SiAlON-SiC composite in the cam/pusher systems does not show these disadvantages: the oscillations and the cyclic variations are much lower. As a result, even if adding silicon carbide to the sialon does not offer the same mechanical efficiency than sialon with a phase-ratio of 60/40 (between 0.9 and 0.95 for the composite and more than 0.95 for sialon with a phase ratio of 60/40), it provides a significant improvement regarding noise emission and cyclic variations.

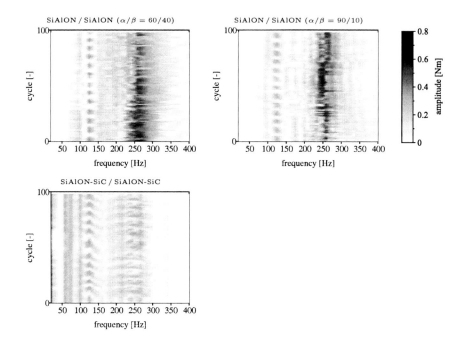

Figure 7.27: Spectrum analysis of the pump driving torque over 100 revolutions with sialon in the cam/pusher system (fuel: gasoline; rotation speed: 500 rpm; delivery pressure: 80 MPa)

As mentioned in subsection 7.2.2, the self-mated silicon nitride pairs could not be tested in the same conditions than the other material pairs. Consequently, no measurement could be performed at 500 rpm and 80 MPa. Figure 7.28 provides the calculated frequency spectra of the silicon nitride pairs at 500 rpm and 50 MPa. The oscillation amplitude scale has been adjusted in order to show amplitudes in the range of 0 % to 10 % of the average pump driving torque. The results indicate high oscillations around 250 Hz with amplitudes over 0.6 N·m (the white spots around 250 Hz correspond to a saturation of the gray-scale). These spectra illustrate the high noise emission, which occurred during the investigations at the test bench. The material pair containing carbon nanotubes shows oscillation amplitudes which are slightly higher. Both material pairs show strong fluctuations of the oscillation amplitude around 250 Hz (between 0.2 N·m and more than 0.6 N·m). However, since the stress cycles and stress duration applied to the self-mated silicon nitride pairs were not the same, no conclusion can be inferred regarding the influence of the CNTs on the noise emission or cyclic variations.

In contrast to the self-mated silicon nitride pairs (with or without CNTs), the combination of silicon nitride with AISI 52100 enabled investigations under the same conditions than the other material pairs (test cycle, test duration). Figure 7.29 illustrates the improved performance in comparison to the self-mated sil-

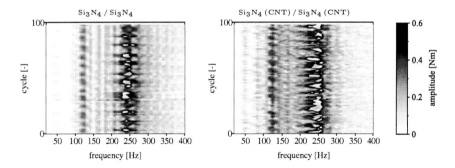

Figure 7.28: Spectrum analysis of the pump driving torque over 100 revolutions with self-mated silicon nitride in the cam/pusher system (fuel: gasoline; rotation speed: 500 rpm; delivery pressure: 50 MPa)

icon nitride pairs. The spectra shown in the figure have been obtained at 500 rpm and 80 MPa delivery pressure after a sliding distance of more than 5,000 m. The oscillation level obtained without CNTs at ca. 250 Hz is comparable to the results obtained with self-mated sialon. However, the cyclic variations with this pair are much lower and similar to the silicon carbide pairs (periodic fluctuations only, no significant frequency shifts). The silicon nitride pair containing CNTs combined with AISI 52100 shows very good results. The amplitude of the oscillations and the cyclic variations are comparable to the results obtained with silicon carbide pairs.

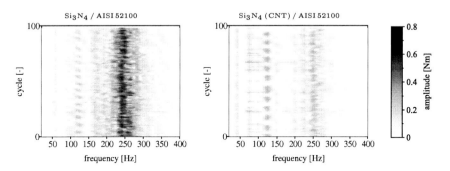

Figure 7.29: Spectrum analysis of the pump driving torque over 100 revolutions with silicon nitride in the cam/pusher system (fuel: gasoline; rotation speed: 500 rpm; delivery pressure: 80 MPa)

7.2.4 Evolution of the efficiencies over time

The results presented in subsection 7.2.2 enable to compare the performance of various material pairs in the cam/pusher systems after a run-in of ca. 5,000 m.

However, the evolution of the mechanical efficiency differs depending on which material combination is considered. Figure 7.30 shows the mechanical efficiency obtained with several material pairs during the run-in period. The operating point represented in this figure is 500 rpm pump rotation speed and 50 MPa delivery pressure. Some of the investigated material pairs are not displayed in the figure since their run-in has been performed with ethanol instead of gasoline.

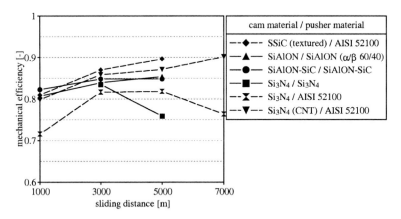

Figure 7.30: Evolution of the pump mechanical efficiency over time with gasoline (rotation speed: 500 rpm; delivery pressure: 50 MPa)

It can be noticed that the mechanical efficiency increases with increasing sliding distance for all the material pairs excepted for the self-mated silicon nitride and the silicon nitride combined with AISI 52100. This improvement is linked to the decreasing friction coefficient observed with the single-piston pump (see figure 7.13) during the run-in period. The silicon nitride combined with AISI 52100 shows decreasing performances after 5,000 m sliding distance whereas the same pair containing CNTs still improves after this distance. An explanation to this observation is provided in subsection 7.2.5.

Figure 7.31 shows the evolution of the mechanical, volumetric and total efficiency measured with silicon carbide combined with AISI 52100 in the cam/pusher system. The operating point considered is 500 rpm and 50 MPa delivery pressure. It can be noticed that the volumetric efficiency of the pump remains approximately constant over the whole testing duration. Consequently, the slightly increase of the pump mechanical efficiency leads to an improvement of the total pump efficiency (from 0.55 at test begin to 0.63 after 5,000 m sliding distance). These results should be compared to the evolution of the commercially available pumps performances at the same operating point (see subsection 7.3.2).

7.2.5 Surface analysis

Analyzing the surfaces of the investigated components enables to complete the information provided by the results presented in the previous subsection. Figure 7.32

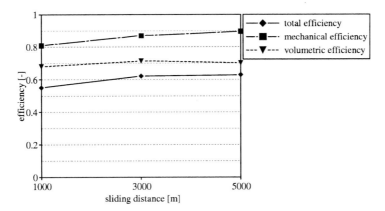

Figure 7.31: Evolution of the total efficiency, mechanical efficiency and volumetric efficiency with gasoline over time (rotation speed: 500 rpm; delivery pressure: 50 MPa; cam material: SSiC (textured), pusher material: AISI 52100)

shows the example of a cam surface after investigation in the 3-piston prototype pump. The bright oblong region in the picture corresponds to the surface in contact with the pusher. Since the investigated materials have a relatively high hardness and since the surfaces are not perfectly planar, more or less smoothened regions can be observed (brighter or darker regions in the picture). The SEM pictures shown in this subsection have all been taken in the center of the cam. The wear is the highest in this region for all the investigated cams since this part of the surface is in permanent contact with the pusher.

Figure 7.32: Overview of a silicon carbide cam after 16 hours of operation (sliding distance: 7,200 m)

Figure 7.33 provides SEM pictures from the silicon carbide cams before and after investigation in the prototype pump. These cams have been tested in self-combination or combined with AISI 52100. The self-mated cams (textured or not) show the same evolution than observed with the single-piston pump: the surfaces are smoothened but no sign of surface damage can be identified. The cavities visible on the surface enable to store fuel for a better lubrication of the sliding system, thus contributing to the good performances observed with self-mated silicon carbide. The texture enhances the lubrication with a higher storage volume for the lubricant and wear particles. The cam used in combination with AISI 52100 illustrates this storage function since most of the micro-dimples in the center of the cam are filled with wear material. An energy dispersive X-ray spectroscopy (EDX) analysis enabled to identify iron as the main component of this wear material. No iron-based layer has been found on the surface excepted in the texture. The first hours of test with this material pair have shown an increasing mechanical efficiency over time followed by stable performance in the last hours. It can be inferred that the micro-dimples were filled during the first hours of the test only, thus avoiding the adhesive action of the steel wear particles released during the run-in period. The surface of the textured silicon carbide cam used in combination with AISI 52100 is smoothened after 7,200 m sliding distance, thus confirming the results obtained with the single-piston pump regarding possible tribochemical reactions in the sliding system.

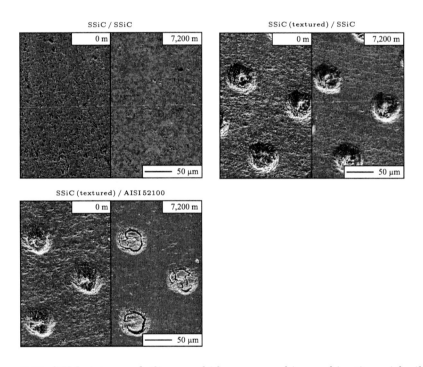

Figure 7.33: SEM pictures of silicon carbide cams used in combination with silicon carbide or AISI 52100 before and after test (sliding distance: 7,200 m)

The surfaces of the sialon cams are presented in figure 7.34. The picture of the sialon cam with an α/β-phase ratio of 60/40 and the picture of the SiAlON-SiC composite lead to the same conclusions than in subsection 7.1.3: the surface is smoothened but some grooves left by the grinding process are still visible. No cavities are visible on the surface of the sialon cams. This may explain the higher cyclic variations and lower mechanical efficiency in comparison to the silicon carbide pairs (since cavities may support the lubrication of the system). The SEM pictures of the sialon cam with an α/β-phase ratio of 90/10 show cracks on the surface with a length of up to 0.5 mm. These cracks indicate a fatigue mechanism occurring on the surface of the cam (see section 4.4). The high torque oscillations amplitudes observed with this material pair are probably one of the main cause which led to this damage. No crack has been found on the cam surfaces with a higher β-phase ratio. This can be explained by the higher fracture toughness of the β-phase in comparison to the α-phase. In addition, the β-phase provides a higher thermal conductivity, thus improving the friction energy dissipation in the pump. Another wear mechanism has been observed with the sialon pairs (excepted the SiAlON-SiC composite). Scratches on the surface of the investigated cams and pushers are visible at macroscopic scale (especially on cams with an α/β-phase ratio of 90/10) as exemplary shown in figure 7.35. These scratches are caused by the abrasive action of wear particles, which cannot be stored in the contact (no cavities or texture) and probably contribute to the cyclic variations observed with pure sialon pairs. The cracks observed on the surface of sialon cams with an α/β-phase ratio of 90/10 may be the cause of a higher particle release in the system, thus leading to more scratches when compared to sialon with an α/β-phase ratio of 60/40. The SiAlON-SiC composite, which is harder and provides a higher thermal conductivity, does not show any scratches of this type on its surface after investigation.

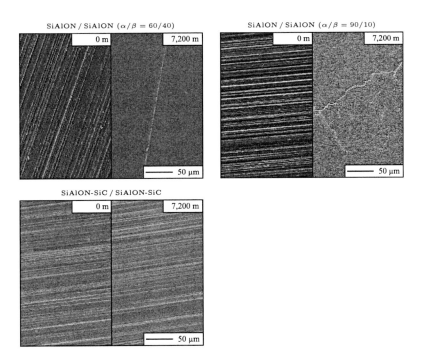

Figure 7.34: SEM pictures of various sialon cams used in self-combination

Figure 7.35: Overview of a sialon cam ($\alpha/\beta = 90/10$) after 16 hours of operation (sliding distance: 7,200 m)

Figure 7.36 depicts silicon nitride cams before and after investigation against pushers made of silicon nitride or AISI 52100. The silicon nitride cams without CNTs show an evolution of their surface which is comparable to the evolution observed with silicon carbide: the surfaces are smoothened but no damages could be identified on the surface after investigation. Like the cams made of sialon, the silicon nitride surfaces (without CNTs) do not show any cavities supporting the lubrication of the sliding system, thus explaining the lower mechanical efficiency in comparison to the other investigated material pairs. Significant differences can be noticed with the silicon nitride cams containing CNTs in comparison to its base material: break-outs are visible on the surface after the grinding process. The lower mechanical properties of the material containing CNTs (especially its lower fracture toughness) may explain this observation. Even if the break-outs on the surface could be considered as a material failure, they may lead to benefits for the present application: the cavities store the lubricant and the wear particles like the texture used with silicon carbide. However, only the cam used in combination with AISI 52100 has shown improved performances (very good mechanical efficiency, low cyclic variations and noise emission). Consequently, it can be inferred that specific tribochemical reactions occur between silicon nitride containing CNTs and AISI 52100.

The SEM pictures of the AISI 52100 pushers used against silicon nitride containing CNTs (see figure 7.37) tend to confirm this supposition. Among the three pushers used for the test in the 3-piston prototype pump, all show particle agglomerates distributed on their surface, especially in the region of the centre (subjected to the highest tribological stresses). According to the EDX analysis, the particles contain chrome and carbon. Consequently, it can be supposed that these particles consist in chromium carbide which is already in the material matrix before testing. None of the pushers made of AISI 52100 used against other materials or pure silicon nitride (without CNTs) shows this type of agglomerates on its surface. It can be assumed that a chemical reaction implicating the iron on the surface of the pushers and the carbon nanotubes contained in the cams occur and reveal the chromium carbide from the AISI 52100 matrix.

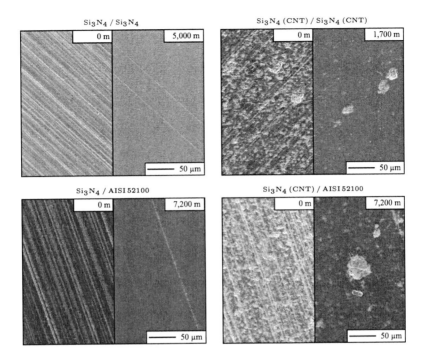

Figure 7.36: SEM pictures of various silicon nitride cams used in self-combination or in combination with AISI 52100

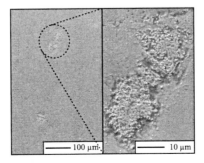

Figure 7.37: SEM pictures of an AISI 52100 pusher used in combination with a silicon nitride cam with carbon nanotubes

7.3 Performance of commercially available pumps

Investigations have been performed with two commercially available pumps (Bosch HDP1 and Bosch CP1) in order to better evaluate the performance of the 3-piston prototype developed as part of this study. The test cycles used for these two pumps were shorter than for the 3-piston prototype pump. The HDP1 has been tested in a restricted test cycle over four hours (corresponding to 320,000 revolutions or 2,560 m sliding distance in the sliding systems) between 10 MPa and 30 MPa (its maximum delivery pressure according to the manufacturer is 20 MPa and external leakage was observed at pressures above 30 MPa). The CP1 has been investigated between 20 MPa and 80 MPa like the 3-piston prototype pump. However, the test had to be interrupted after eight hours (400,000 revolutions or 5,600 m sliding distance) because of the decreasing and irregular performance of the pump. Unless mentioned otherwise, the results shown in this section have been obtained in the last test cycle.

7.3.1 Mechanical efficiency and cyclic variations

Since the production pumps are not fitted with cylinder pressure transducers, the mechanical efficiency could not be calculated via the pump mean effective pressure as described in subsection 7.2.2. Consequently the mechanical η_{mech} efficiency has been calculated via equation 3.21 by neglecting the hydraulic losses:

$$\eta_{mech} = \frac{\eta_{total}}{\eta_{vol}} \tag{7.4}$$

The method used for the frequency analysis of the pump driving torque is slightly different from the method used in subsection 7.2.3. Since the cylinder pressure traces were not available, the theoretical pump driving torque could not be calculated. As a result, the theoretical frequency spectrum of the driving torque (assuming that no friction occurs in the pump) could not be subtracted to the analysis of the real data. This means that the spectra shown in figures 7.38 and 7.39 include all the torque oscillations (the results corresponding to the 3-piston prototype pump only included the oscillations caused by friction).

Figure 7.38 shows the mechanical efficiency and the torque spectrum analysis corresponding to the Bosch HDP1 gasoline pump. The mechanical efficiency is in the range of 0.3 to 0.5 over the entire investigated speed and pressure range. In comparison, the mechanical efficiency of the prototype pump between 20 MPa and 30 MPa was higher than 0.6 with all the investigated material pairs. Delivery pressure levels above 20 MPa could not be reached at 300 rpm because the volumetric losses were too high ($\eta_{vol} = 0$). Measurements at 30 MPa and rotation speeds above 900 rpm were not possible because one of the pistons jammed at TDC.

The torque frequency analysis at 20 MPa and 500 rpm shows stable operation with low cyclic variations. Periodical fluctuations of the oscillation amplitude can be observed like with silicon carbide pairs in the prototype pump. The high amplitude of the low frequencies (50 Hz and below) are essentially caused by the normal operation of the pump and does not indicate any high friction level in this frequency range.

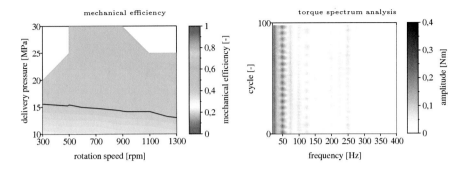

Figure 7.38: Mechanical efficiency and torque spectrum analysis at 20 MPa and 500 rpm of the Bosch HDP1 (delivered fuel: gasoline)

The mechanical efficiency and the torque spectrum analysis of the Bosch CP1 is shown in figure 7.39. Even if the CP1 is capable of delivering Diesel fuel at pressures of up to 135 MPa, it was not possible to perform measurements in the entire speed and pressure range. Several operating points at high delivery pressure and low rotation speed could not be investigated because of the very high volumetric losses (the desired pressure level could not be reached). At high rotation speeds, the pistons jammed regularly in the cylinders and did not move to the BDC, thus causing high and irregular volumetric losses. Consequently, several operating points have not been investigated at 1,300 rpm. At low rotation speed and delivery pressures below 50 MPa, the CP1 has a mechanical efficiency of approximately 0.5. Its performance improves with increasing speed and the mechanical efficiency reaches ca. 0.9.

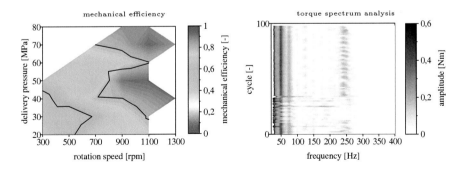

Figure 7.39: Mechanical efficiency and torque spectrum analysis at 50 MPa and 500 rpm of the Bosch CP1 (delivered fuel: gasoline)

Figure 7.39 enables to visualize the high cyclic variations of the driving torque over time (amplitude fluctuations and frequency shifts). These variations correspond to the phenomenon previously mentioned: some pistons jam in the cylinder and do not move to BDC during a few driveshaft revolutions, thus reducing the pump volumetric efficiency on the one hand and causing high torque variations on the other hand.

7.3.2 Evolution of the Bosch CP1 efficiency over time

During the investigations, an increasing driving torque and a decreasing pump mass flow has been observed with the Bosch CP1 at a single operating point. Figure 7.40 shows the evolution of the pump efficiencies over time measured at 500 rpm rotation speed and 50 MPa delivery pressure. In contrast to the results obtained with the 3-piston prototype pump (see figure 7.31), the volumetric efficiency decreases over time. Even if the mechanical efficiency slightly increases, the volumetric losses cause the reduction of the pump total efficiency by 30 % within eight hours. This can be explained by several effects. The high friction levels in the sliding systems cause a reduced motion of the piston as already mentioned in subsection 7.3.1. This phenomenon leads to a lower pump mass flow. Moreover, the wear in the piston/cylinder system may lead to an increasing gap between the two components, thus causing higher internal leakage.

In contrast to the Bosch CP1, the Bosch HDP1 has only been tested during a short time. Consequently, comparing the evolution of its efficiencies with the other investigated pumps is not meaningful.

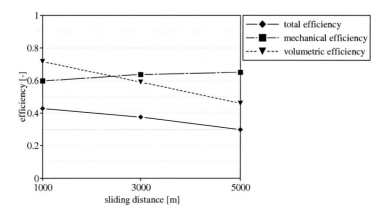

Figure 7.40: Total pump efficiency and mechanical efficiency of the Bosch CP1 over time (rotation speed: 500 rpm; delivery pressure: 50 MPa)

7.3.3 Wear analysis

The production pumps were disassembled after the experiments on the test bench in order to evaluate the wear of their components. The damages on the surfaces can be observed at macroscopic scale. Since the HDP1 has only been tested shortly above its nominal delivery pressure, no significant wear is visible after investigation excepted on the surface of the piston: a part of the coating on the contact surface disappeared (see picture in figure 7.41).

Figure 7.41: Pictures of a Bosch HDP1 piston after 4 hours of investigation (sliding distance: 4,200 m)

The pictures of a cam, pusher and piston of the Bosch CP1 (figure 7.42) support the inference from the previous subsection: the coating of the cam disappeared in the stressed region and has been partially transferred to the surface of the pusher. Furthermore, scratches are visible on the surface of the pushers and pistons (despite a hardness of ca. 720 HV and 700 HV for the pushers and pistons, respectively), thus highlighting the high friction levels in both sliding systems. These surface damages, which appeared within a few hours, explain the decreasing volumetric efficiency over time. Furthermore, it indicates that conventional materials do not provide sufficient wear resistance for pump delivery pressures of up to 80 MPa with fuel lubrication.

Figure 7.42: Pictures of a Bosch CP1 cam (left), pusher (middle) and piston (right) after investigation (sliding distance: 5,600 m)

8 Discussion

The results presented in the previous chapter provide information about the performance of several material combinations in the sliding systems of gasoline pumps. Each material combination offers specific advantages and disadvantages. This chapter summarizes some of the main criteria which should help to select appropriate material pairs that enable to increase the delivery pressure of fuel-lubricated high-pressure gasoline pumps. In addition, the influence of high-pressure levels on the pump volumetric efficiency and the impact on the engine performances are discussed in this chapter.

8.1 Appropriate material pairs for the application in gasoline-lubricated sliding systems

The friction losses and the wear resistance of a material combination are two critical criteria for the application in highly stressed tribotechnical systems. Stable performance, low noise emission and costs are further key criteria.

8.1.1 Friction losses

The friction losses in the sliding systems affect the pump mechanical efficiency. Consequently, higher friction losses cause an increase of the pump driving torque, thus impacting the performance of the combustion engine (see section 8.4). High friction levels may also lead to piston jamming in the cylinders and cause a reduced volumetric efficiency (as observed with the Bosch CP1).

Depending on the operating point considered, several material combinations in the cam/pusher system show very good performances. By considering the best operating point of each material pair, the best results are obtained with self-mated silicon carbide, self-mated sialon (with an α/β-phase ratio of 60/40) and CNTs-containing silicon nitride combined with hardened bearing steel (AISI 52100). These material pairs have shown a mechanical efficiency of more than 0.95 at their best operating point (usually at the highest investigated rotation speed and delivery pressure). Texturing the cam of the self-mated silicon carbide combination enables to extend the high-efficiency range in comparison to the untextured pair. If the low rotation speed and low delivery pressure range is considered, the lowest friction losses were achieved with the self-mated silicon carbide pair. This material pair offers a mechanical efficiency of more than 0.75 in the entire investigated operating range.

Some of the investigated material pairs are not appropriate for the application in highly-stressed sliding systems regarding friction losses. The combination of silicon carbide with hardened bearing steel leads to adhesion and shows an increasing friction coefficient over time. However, texturing the silicon carbide cam enables to overcome this problem and leads to satisfactory performances regarding the pump mechanical efficiency. Self-mated sialon with an α/β-phase ratio of 90/10, self-mated silicon nitride (with or without CNTs) and silicon nitride (without CNTs) combined with hardened bearing steel show high friction levels in the low speed range (pump mechanical efficiency of 0.6 or less). Moreover, these pairs reach a maximum mechanical efficiency which is lower than the best pairs (lower than 0.9 at the best operating point).

Self-mated silicon carbide and silicon carbide in combination with hardened bearing steel both offer good performances in the piston/cylinder system (similar work of friction). Consequently, other criteria may help to select the ideal pair for this sliding system.

8.1.2 Wear acceptance

As mentioned in chapter 4, wear may lead to surface damages and failure of the tribotechnical system. Moreover, the progressive material loss on the surface of the bodies may affect the hydraulic performance of the pump. Material loss in the piston/cylinder system leads to an increase of the throttle seal gap and thus cause higher internal leakage (reduction of the pump volumetric efficiency). Wear in the cam/pusher system changes the position of the piston at TDC and increases the dead volume in the working chambers. A larger dead volume contributes to reduce the pump volumetric efficiency. Consequently, high wear rates in the sliding systems of a high-pressure pump may lead to a decreased volumetric performance.

Cam/pusher system
None of the investigated material pairs in the cam/pusher system have shown significant material loss after testing in any of the prototype pumps. Only a smoothing of the surface asperities could be observed. The specific wear rate could be estimated for self-mated sialon only (below $1.6 \cdot 10^{-8} \frac{mm^3}{N \cdot m}$) and is considerably lower than usual wear rates tolerated in other tribotechnical systems (see figure 4.8). The impact of a wear rate W_{rate} of $1.6 \cdot 10^{-8} \frac{mm^3}{N \cdot m}$ in the cam/pusher system on the volumetric performance of the pump can be evaluated with following assumptions:

- the pump always delivers the fuel at its maximum delivery pressure p_{high} of 80 MPa

- the average pressure $p_{average}$ in the work chamber at 80 MPa delivery pressure is 38 MPa (calculated via measured data)

- the delivery pressure of the feed pump p_{low} is 0.5 MPa

- the isothermal compressibility factor of gasoline is $\beta_T = 110 \cdot 10^{-6} \frac{1}{bar}$

- the piston diameter d_{piston} is 8 mm, the piston travel T_{piston} is 4 mm and the pusher diameter d_{pusher} is 20 mm (the pump displacement V_H is 603 mm^3)

- the cam and the pusher are made of the same material (same wear rate)

- the worn surface of the cam is approximated by a circular area with a diameter d_{cam} of 20 mm

- the friction forces between the pistons and cylinders are neglected (average value of approximately 0 N over one driveshaft revolution, according to measured data on the single-piston prototype pump)

- the average force F_{spring} applied by the retaining springs to the pusher is 40 N

- the average combustion engine rotation speed is 2,000 rpm

- the high-pressure pump is driven by the camshaft of the engine (average pump rotation speed N_{pump} of 1,000 rpm)

- the average speed of the passenger car V_{car} is 50 km/h

- the car travels a distance S_{car} of 250,000 km during its lifetime

The average force F_{cam} applied to the cams under these conditions is given by:

$$F_{cam} = \frac{d_{piston}^2 \cdot \pi}{4} \cdot p_{average} + F_{spring} = 1950 \, \text{N} \qquad (8.1)$$

The number of revolutions R_{pump} achieved during the pump lifetime is:

$$R_{pump} = N_{pump} \cdot \frac{60}{V_{car}} \cdot S_{car} = 3 \cdot 10^8 \qquad (8.2)$$

The sliding distance S_{cam} in the cam/pusher system can be calculated as follows:

$$S_{cam} = 2 \cdot R_{pump} \cdot T_{piston} = 2.4 \cdot 10^6 \, \text{m} \qquad (8.3)$$

Following equation provides the wear volume V_{wear} for each bodies of the cam/pusher system:

$$V_{wear} = W_{rate} \cdot S_{cam} \cdot F_{cam} \approx 75 \, \text{mm}^3 \qquad (8.4)$$

The wear volume leads to the gap $\triangle TDC$ between the initial position of the piston at TDC and the position at the end of the lifecycle (assuming that the cam and the pusher are subjected to the same wear rate):

$$\triangle TDC = 2 \cdot \frac{V_{wear}}{\frac{d_{pusher}^2 \cdot \pi}{4}} \approx 0.48 \, \text{mm} \qquad (8.5)$$

This gap causes an increase of the dead volume $\triangle V_c$ in each cylinder as follows:

$$\triangle V_c = \triangle TDC \cdot \frac{d_{piston}^2 \cdot \pi}{4} \approx 24 \, \text{mm}^3 \tag{8.6}$$

The reduction of the suction volume $\triangle V_s$ in each cylinder resulting from the increased dead volume can be calculated via equation 3.17:

$$\triangle V_s = \frac{\triangle V_c}{\frac{1}{\beta_T \cdot (p_{high} - p_{low})} - 1} = 2.3 \, \text{mm}^3 \tag{8.7}$$

As a result, $3 \cdot \triangle V_s = 6.9 \, \text{mm}^3$ are lost by each revolution of the 3-piston pump at the end of its lifecycle. Assuming that there is no initial dead volume in the pump and that no leakage occurs between the piston and the cylinder, the pump mass flow is reduced by $\frac{3 \cdot \triangle V_s}{V_H} \cdot 100 \approx 1.1\%$ at the end of its lifecycle because of the material loss in the cam/pusher systems. This decreased mass flow should be taken into account for the pump design but does not represent any other major difficulty. Moreover, this exemplary calculation is performed with a wear rate of $1.6 \cdot 10^{-8} \, \frac{\text{mm}^3}{\text{N} \cdot \text{m}}$. This value is higher than the actual wear rate corresponding to the self-mated sialon pair (see subsection 7.1.3 for more details). Consequently, the real impact of wear in the cam/pusher system with self-mated sialon is even lower.

Some of the investigated material pairs in the cam/pusher system have shown wear mechanisms which could lead to pump failure. Consequently, silicon carbide combined with hardened bearing steel (adhesion mechanism with a steel layer visible on the ceramic cam after testing) or self-mated sialon with a α/β-phase ratio of 90/10 (cracks) are not appropriate for this application regarding wear.

Piston/cylinder system
No wear could be measured in the piston/cylinder sliding systems. Consequently, estimating the increase of internal leakage over the pump lifecycle on the basis of measured data is not possible. Since the leakage mass flow increases with the cube of the throttle gap (see equation 3.16), material losses in this system may affect the internal leakage dramatically. No significant decrease of the pump volumetric efficiency could be observed during investigation with a single set of pistons and cylinders. In contrast, the effect of fluctuating dimensions (within the dimensional tolerances) between different set of pistons and cylinders was clearly noticeable. This phenomenon is discussed in section 8.3.

8.1.3 Noise emission and cyclic variations

The noise emission and the cyclic variations evaluated in subsection 7.2.3 are the consequence of friction and wear phenomenon in the sliding systems (especially in the cam/pusher system, which is subjected to the highest tribological stresses). The cyclic variations highlight irregular friction mechanisms, probably caused by the intermittent abrasive effect of wear particles released in the contact. Consequently, high cyclic variations should be prevented for the application in a high-pressure pump. High noise emission levels do not mean that the material pair is not appropriate for application regarding friction or wear. However, since noise emission is related to comfort, a low level should be preferred.

All the investigated material pairs have shown good performances regarding cyclic variations excepted the self-mated sialon with an α/β-phase ratio of 60/40 and 90/10. The self-mated silicon nitride pairs have shown high torque oscillations in the audible frequency range. Self-mated sialon pairs (excepted SiAlON-SiC composite) also caused noise emission during investigation but at a lower level.

8.1.4 Design complexity and costs

Giving a quantitative estimation of the manufacturing costs of each component is not meaningful since the costs are strongly influenced by the technical specifications of the pump and the economies of scale, which could be realized via series production. However, a qualitative evaluation of the manufacturing costs is possible by considering the design complexity. The use of ceramic components requires some specific design methods for an efficient and reliable integration in the pump. Moreover, the shaping of ceramic parts is relatively expensive and simple geometries should be privileged. Consequently, it can be considered that limiting the number of ceramic components in the sliding systems of a high-pressure pump is an advantage regarding design complexity, constructed size and manufacturing costs. The combination of ceramic parts with hardened bearing steel offers the benefit of improved performances (in comparison to conventional materials) and a limited use of ceramic components. Two material pairs including pushers made of steel have shown satisfactory performances: textured silicon carbide and silicon nitride containing carbon nanotubes. Concerning the piston/cylinder system, since the two investigated material pairs have shown similar performance, the combination of steel pistons with cylinders made of silicon carbide should be preferred regarding design complexity and costs.

8.1.5 Qualitative evaluation of the investigated material pairs in the cam/pusher system

Table 8.1 provides a qualitative evaluation of the material pairs tested in the cam/pusher systems regarding friction losses, wear resistance, noise and cyclic variations and design complexity. This evaluation is not meant to indicate which material combination best fits for an application in a fuel-lubricated high-pressure gasoline pump since the criteria should be weighted according to a given specifications sheet. The mark "+" indicates that the material pair is notably appropriate for the application according to the criterion considered. In contrast, the mark "-" means that the material pair is not appropriate. The "o" indicates that even if the material combination does not offer the best properties according to the criterion considered, it could be used anyway with other benefits. For example, the SiAlON-SiC composite does not show the best performances regarding friction losses but provides stable performances in contrast to the combination of silicon carbide (without texture) with AISI 52100. The self-mated silicon nitride pairs have no mark regarding wear resistance because they have not been tested under the same conditions than the other investigated materials. As a result, no conclusions can be drawn about their performance regarding this criterion. The material pairs with silicon nitride containing CNTs have been marked "o" since break-outs were visible on the surface after the grinding process (before testing). Even if these break-outs did not lead to damages during the investigations, it indicates a material weakness, which may impact the performances over a longer operating duration. Furthermore, the possible tribochemical reaction between steel and the silicon nitride containing CNTs requires further investigation in order to ensure that no severe wear may occur after a longer sliding distance. It can be assumed that the use of a ceramic part in the cam/pusher system leads to higher costs caused by manufacturing on the one hand and by the integration into the system on the other hand. Estimating the cost of the material itself is not meaningful as several materials used for investigation are at the research stage. Consequently, a bonus mark has been accorded to ceramic/steel combinations whereas the self-mated ceramic pairs have been marked "o".

The qualitative evaluation shown in table 8.1 enables to notice that several material pairs could be used in the cam/pusher systems of a fuel-lubricated high-pressure gasoline pump. Self-mated silicon carbide (textured or not), textured silicon carbide combined with AISI 52100 or silicon nitride containing CNTs combined with AISI 52100 offer very good global performances. In contrast, silicon carbide (without texture) combined with AISI 52100, self-mated sialon with an α/β-phase ratio of 90/10, self mated silicon nitride or silicon nitride (without CNTs) combined to AISI 52100 are not appropriate for this application.

Material combination	Criterion			
	Friction losses	Wear resistance	Noise and cyclic variations	Design complexity
SSiC / SSiC	+	+	+	o
SSiC (textured) / SSiC	+	+	+	o
SSiC / AISI 52100	-	-	not measured	+
SSiC (textured) / AISI 52100	+	+	+	+
SiAlON /SiAlON (60/40)	+	+	o	o
SiAlON / SiAlON (90/10)	-	-	o	o
SiAlON-SiC / SiAlON-SiC	o	+	+	o
Si_3N_4 / Si_3N_4	-	?	-	o
Si_3N_4 (CNT) / Si_3N_4 (CNT)	-	?	-	o
Si_3N_4 / AISI 52100	-	o	o	+
Si_3N_4 (CNT) / AISI 52100	+	o	+	+

Table 8.1: Evaluation of the investigated material pairs in the cam/pusher system

8.2 Optimization of the pump specifications for an improved mechanical efficiency

Some investigated material pairs have shown high friction losses at low rotation speed when compared to the best combinations, but satisfactory results at higher speed. The properties of the 3-piston prototype pump imply low relative sliding speeds in its sliding systems. Increasing the sliding speed and thus reduce the average friction losses is possible by increasing the piston travel (via the driveshaft eccentricity). In order to keep a constant displacement, such a solution requires to reduce the piston diameter. However, manufacturing inaccuracies and the deformation of the piston during operation may lead to a lower limit of the piston diameter. Consequently, it is necessary to make a compromise between pump efficiency and manufacturing costs.

Material combinations such as the SiAlON-SiC composite have shown an optimal operating range at a pump delivery pressure below the maximum investigated level. The optimal delivery pressure can be shifted in the pump operating map by changing the ratio between the piston diameter and the pusher diameter. A larger pusher (with an unchanged piston diameter) enables to reduce the surface pressure on the cam and thus to increase the delivery pressure at constant mechanical efficiency. However, increasing the diameter of the pusher causes a larger constructed size and may lead to conflicts regarding pump integration in a passenger car. A reduced piston diameter (pusher diameter unchanged) combined with an increased piston travel also enables to reduce the force applied to the cam. However, the difficulties previously mentioned about small piston diameters should be taken into account.

8.3 Volumetric losses

The investigations presented in this document focus on the friction losses and wear caused by high tribological stresses in the sliding systems. However, increasing the delivery pressure of gasoline high-pressure pumps also affects the pump volumetric efficiency considerably. The increasing volumetric losses are mainly caused by three effects. First, the internal leakage between pistons and cylinders reduce the delivered fuel through the pressure valve. In addition, the re-expansion in the dead volumes lead to a reduced suction. Lastly, the elastic deformation of the pump causes an increase of the dead volumes.

The absolute internal leakage is barely affected by the pump rotation speed at constant delivery pressure. This means that the ratio between the internal leakage and the pump mass flow decreases with increasing pump rotation speed. The upper limit of the throttle gap size is given by the maximum leakage flow tolerated at low rotation speed. A small throttle gap is preferable in order to reduce the internal leakage. However, manufacturing accuracy and possible deformations (mechanically or thermally induced) lead to a lower limit of the throttle gap size. The nominal throttle gap between the pistons and the cylinders in the 3-piston prototype pump is 5 μm. The corresponding dimensional tolerance is ±1 μm. The relative leakage mass flow at 80 MPa delivery pressure can be calculated with following assumptions via equation 3.16:

- the average pressure in the working chamber at 80 MPa delivery pressure is 38 MPa

- the delivery pressure of the feed pump is 0.5 MPa

- the nominal diameter of the piston and cylinder is 8 mm

- the pump displacement is 603 mm^3 (corresponding to the displacement of the 3-piston prototype pump)

- the length of the throttle gap seal is 16 mm

- the dynamic viscosity of gasoline is 0.65 mPa·s

- the density of gasoline is 750 $\frac{\text{kg}}{\text{m}^3}$

Figure 8.1 shows the theoretical relative internal leakage at 80 MPa delivery pressure as a function of the pump rotation speed between 300 rpm and 3,500 rpm (calculated via equation 3.16). In this example, the fuel is considered as being incompressible (no re-expansion losses). The hatched area indicates the relative internal leakage range within the dimensional tolerance of the throttle seal gap. At 300 rpm, the fluctuation of the leakage mass flow is between ca. 20 % and 95 % of the theoretical pump mass flow. At nominal throttle gap size (5 μm), the internal leakage is approximately 50 % of the theoretical pump mass flow. The leakage ratio decreases with increasing speed and is below 20 % at rotation speeds higher than 1,500 rpm. This diagram shows that even very small dimensional fluctuations lead to very high leakage flow variations at low pump rotation speed. Since no significant variation of the pump delivery mass flow could be identified during the investigations with a single set of pistons and cylinders, it can be inferred that the dimensional tolerance of the components have a higher influence on the leakage mass flow than the wear of the investigated material pairs in the piston/cylinder systems.

The share of volumetric losses caused by re-expansion in the dead volumes is constant over the entire speed range and increases with increasing delivery pressure (see equation 3.17). The influence of the dimensional tolerances or wear on the dead volume is considerably lower than the influence of the throttle gap size on the internal leakage. As exemplary calculated for the 3-piston prototype pump in subsection 8.1.2, shifting the TDC by 0.48 mm causes a decreased delivery mass flow by ca. 1.1 %. Furthermore, the use of ceramic components in the sliding systems offer a higher rigidity in comparison to metallic materials, thus limiting the elastic deformation caused by the pressure forces.

An example of volumetric efficiency measured with the 3-piston prototype pump at 500 rpm is shown in figure 8.2. The pressure range is 20 MPa to 80 MPa. The diagram shows the results obtained with gasoline and ethanol with a single set

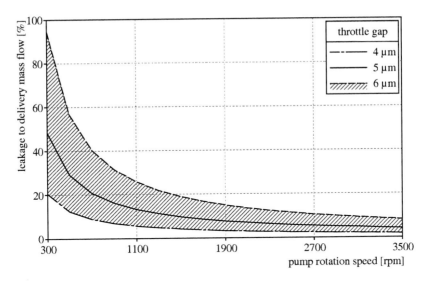

Figure 8.1: Internal leakage to theoretical delivery mass flow ratio as a function of the throttle gap at 80 MPa delivery pressure

of pistons and cylinders. The volumetric efficiency with gasoline decreases from 0.82 to 0.38 between 20 MPa and 80 MPa delivery pressure. Ethanol causes lower volumetric losses (especially at high delivery pressures) since it provides a higher viscosity, thus limiting the internal leakage (see equation 3.16). These results show that the volumetric losses have a higher influence on the total pump efficiency than the mechanical losses with appropriate ceramic material pairs in the sliding systems.

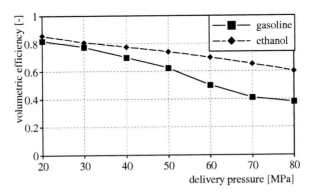

Figure 8.2: Volumetric efficiency with gasoline and ethanol as a function of the delivery pressure at 500 rpm

8.4 Impact on the engine performance

Increasing the delivery pressure of the gasoline pump offers the opportunity to reduce the pollutants emissions (especially the soot emission). However, a higher delivery pressure necessarily implies a higher pump driving power (increase of the pressure forces, mechanical losses and volumetric losses). Since the pump is driven by the internal combustion engine, increasing the pump delivery pressure may cause a higher fuel consumption. However, it has been shown in [44] and [15] that an increase of the fuel injection pressure in stratified mode enables to slightly reduce the specific fuel consumption of the engine. This section aims at evaluating the impact of an increased injection pressure on the fuel consumption of the combustion engine by taking the two previously mentioned effects into account. This evaluation does not provide information over the entire engine operating map since the required data is available at two operating points only.

Figure 8.3 provides the indicated specific fuel consumption (ISFC) measured by Buri and Kneifel in [15] and [44] on a direct-injection engine operated in stratified mode. The diagram shows the ISFC measured at 3 bar and 6 bar Indicated Mean Effective Pressure (IMEP), 2,000 rpm. The experiments were performed at injection pressures of up to 100 MPa but have been restricted to the range of 20 MPa to 80 MPa in the figure in order to fit the investigated pressure range of the 3-piston prototype pump. At 3 bar IMEP, the ISFC decreases from 246 $\frac{g}{kWh}$ at 20 MPa injection pressure to 230 $\frac{g}{kWh}$ at 50 MPa and remains constant when increasing the pressure to 80 MPa. This corresponds to a reduction of the fuel consumption of 6.5 %. At 6 bar IMEP, the ISFC decreases from 222 $\frac{g}{kWh}$ to 218 $\frac{g}{kWh}$ between 20 MPa and 50 MPa injection pressure (-1.8 %) and to 217 $\frac{g}{kWh}$ at 80 MPa injection pressure (-2.25 %).

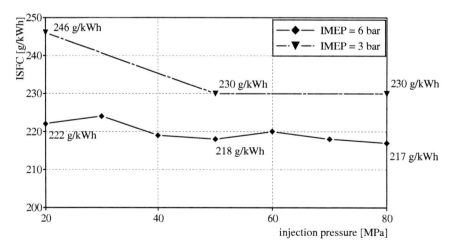

Figure 8.3: Indicated specific fuel consumption as a function of the injection pressure at 2,000 rpm, 3 bar and 6 bar indicated mean effective pressure [15, 44]

The reduced ISFC at higher injection pressure provides an effective power margin which can be used for driving the high-pressure gasoline pump. If this margin is higher than the additional driving power required for the pump, increasing the injection pressure does not lead to an increased fuel consumption and may still provide a fuel economy improvement. If the pump driving power is higher than the calculated effective power margin, increasing the injection pressure causes a higher fuel consumption.

Estimating the effective power margin first requires the calculation of the indicated and effective engine power. The indicated engine power P_i is given as a function of the rotation speed n, the indicated mean effective pressure p_{mi} and the engine displacement V_H ($652\,\mathrm{cm}^3$) by equation 8.8 [53]. The effective engine power $P_{eff,engine}$ can be calculated via equation 8.9 by assuming that the engine mechanical efficiency η_{mech} is 0.75 at 3 bar IMEP and 0.85 at 6 bar IMEP.

$$P_i = 0.5 \cdot n \cdot p_{mi} \cdot V_H \tag{8.8}$$

$$P_{eff,engine} = \eta_{mech} \cdot P_i \tag{8.9}$$

At 2,000 rpm and 3 bar IMEP, the engine effective power is 2.445 kW. At the same rotation speed and 6 bar IMEP, the effective power is 5.542 kW. The effective power margin can be estimated via the relative reduction of the ISFC (in %) and is given in table 8.2 for an injection pressure of 50 MPa and 80 MPa. The reference injection pressure is 20 MPa.

fuel pressure	engine effective power margin (ratio of the engine effective power)	
	IMEP=3 bar	IMEP=6 bar
50 MPa	159 W (6.5 %)	100 W (1.8 %)
80 MPa	159 W (6.5 %)	125 W (2.25 %)

Table 8.2: Engine effective power margin at 2,000 rpm

The fuel mass flow $\dot{m}_{fuel}$ required by the engine at the investigated operating points is given by following equation:

$$\dot{m}_{fuel} = ISFC \cdot P_i \tag{8.10}$$

This fuel mass flow enables to calculate the pump effective power $P_{eff,pump}$ as a function of the injection pressure p_{inj} (it is assumed that the fuel density ρ_{fuel} is $750\,\frac{\mathrm{kg}}{\mathrm{m}^3}$ and that the feed pump delivers the fuel at $p_{low} = 0.5\,\mathrm{MPa}$):

$$P_{eff,pump} = \frac{\dot{m}_{fuel}}{\rho_{fuel}} \cdot (p_{inj} - p_{low}) \tag{8.11}$$

Table 8.3 gives the pump effective power at 3 bar and 6 bar IMEP, 2,000 rpm. Two cases are considered:

- the pump provides an ideally regulated fuel mass flow (only the required fuel mass flow is delivered; the regulation system does not cause any energy losses)

- the pump does not provide any regulated fuel mass flow (the pump always delivers the maximum fuel mass flow)

fuel pressure	pump effective power (additional pump effective power in comparison to 20 MPa fuel pressure)		
	regulated fuel mass flow		no mass flow regulation
	IMEP=3 bar	IMEP=6 bar	
20 MPa	6 W	10 W	28 W
50 MPa	15 W (+9 W)	27 W (+17 W)	72 W (+58 W)
80 MPa	24 W (+18 W)	43 W (+33 W)	115 W (+101 W)

Table 8.3: Pump effective power at 2,000 rpm

With a regulated fuel mass flow, increasing the injection pressure from 20 MPa to 80 MPa leads to an increase of the pump effective power of 18 W at 3 bar IMEP, 2,000 rpm. At 6 bar IMEP, the effective pump power increases by 33 W at 80 MPa injection pressure in comparison to 20 MPa injection pressure. The maximum fuel mass flow has been calculated via following assumptions: the maximum engine load is 12 bar IMEP and the maximum ISFC at full load is 300 $\frac{g}{kWh}$. Consequently, the effective pump power at 2,000 rpm increases by 101 W between 20 MPa and 80 MPa injection pressure (independently from the engine load). The data given in table 8.3 shows that the theoretical effective pump power is lower than the effective power margin provided by the engine with high injection pressure at each investigated operating point (with or without fuel mass flow regulation).

The pump driving power can be estimated via its total efficiency (see equation 3.13). Figure 8.4 shows the measured total efficiency of the investigated 3-piston pumps (prototype pump, Bosch CP1 and Bosch HDP1) in the delivery pressure range of 20 MPa to 80 MPa and at a pump rotation speed of 1,000 rpm (engine rotation speed of 2,000 rpm). The increasing volumetric efficiency of the prototype pump and the CP1 in the lowest pressure range can be explained by the improved mechanical efficiency at high delivery pressure (see chapter 7). However, the volumetric losses outweighs the mechanical performance and cause a decreasing total efficiency in the high-pressure range. These pumps are not designed for the fuel consumption of the combustion engine considered in this section. Moreover, these pumps do not provide any fuel mass flow regulation. Consequently, the calculated pump driving power of each pump is fictive and does not correspond to measured data (their displacement is virtually reduced). However, it enables to estimate the impact of an increased injection pressure on the engine performance with a pump that fits its fuel consumption.

Table 8.4 shows the fictive driving power required by the investigated pumps at 2,000 rpm engine speed and 3 bar IMEP. It is assumed that the pumps provide

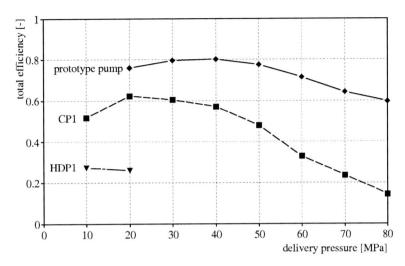

Figure 8.4: Total efficiency of the investigated high-pressure pumps at 1,000 rpm

an ideal mass flow regulation. Due to its low efficiency, the driving power of the HDP1 is significantly higher than the other pumps. Increasing the fuel pressure to 50 MPa leads to a driving power increase of 11 W for the prototype pump and 22 W for the CP1. At 80 MPa fuel pressure, the prototype pump requires 32 W more than at 20 MPa, whereas the CP1 requires and additional driving power of 156 W. Since the HDP1 was not investigated at 50 MPa and more, no estimation of its driving power can be performed. Comparing the results from table 8.4 to the effective power margin of 159 W (see table 8.2) indicates that the reduction of the fuel consumption at 50 MPa and 80 MPa outweighs the increased pump driving torque for the prototype pump and the CP1.

fuel pressure	pump input power		
	(additional input power in comparison to 20 MPa fuel pressure)		
	prototype	CP1	HDP1
20 MPa	8 W	9 W	22 W
50 MPa	19 W (+11 W)	31 W (+22 W)	-
80 MPa	40 W (+32 W)	165 W (+156 W)	-

Table 8.4: Influence of the high-pressure pump driving power on the engine performance at 2,000 rpm, IMEP=3 bar (ideally regulated fuel mass flow)

The fictive driving power of the investigated pumps (with ideal mass flow regulation) at 2,000 rpm engine speed and 6 bar IMEP is given in table 8.5. The required driving power of the prototype pump increases by 17 W at 50 MPa and 57 W at 80 MPa fuel pressure, respectively. Since the effective power margin is 100 W at 50 MPa and 125 W at 80 MPa, the impact of the prototype pump driving power on the fuel consumption is lower than the benefit offered by the lower ISFC. The additional power required by the CP1 at 50 MPa fuel pressure is also below the effective power margin. In contrast, at 80 MPa injection pressure, the fictive driving power of the CP1 is well above the effective power margin of 125 W. This means that increasing the injection pressure to 80 MPa with the CP1 would lead to a fuel penalty.

fuel pressure	pump input power		
	(additional input power in comparison to 20 MPa fuel pressure)		
	prototype	CP1	HDP1
20 MPa	14 W	17 W	40 W
50 MPa	34 W (+17 W)	55 W (+38 W)	-
80 MPa	71 W (+57 W)	297 W (+280 W)	-

Table 8.5: Influence of the high-pressure pump driving power on the engine performance at 2,000 rpm, IMEP=6 bar (ideally regulated fuel mass flow)

Table 8.6 shows the fictive driving power of the investigated fuel pumps without fuel mass flow regulation. Since the pump always delivers the maximum fuel quantity, the driving power is significantly higher than in the case of an ideally regulated mass flow. The prototype pump requires 56 W more at 50 MPa and 156 W more at 80 MPa in comparison to an injection pressure of 20 MPa. Comparing these results to the effective power margin at 3 bar IMEP indicates that increasing the delivery pressure of the prototype pump would not cause an increase of the fuel consumption. At 6 bar IMEP and 50 MPa delivery pressure, the additional power required by the prototype pump is lower than the engine effective power margin. At 80 MPa fuel pressure, the additional driving torque of 156 W slightly outweighs the effective power margin, thus leading to a small fuel penalty. The additional driving power of the CP1 at 50 MPa is approximately equal to the engine effective power margin. At 80 MPa, the CP1 would cause an increase of the fuel consumption since it requires an additional driving power of 658 W.

fuel pressure	pump input power		
	(additional input power in comparison to 20 MPa fuel pressure)		
	prototype	CP1	HDP1
20 MPa	37 W	45 W	108 W
50 MPa	93 W (+56 W)	150 W (+105 W)	-
80 MPa	193 W (+156 W)	803 W (+658 W)	-

Table 8.6: Influence of the high-pressure pump driving power on the engine performance at 2,000 rpm (no fuel mass flow regulation)

The results presented in this section indicate that an increase of the pump delivery pressure does not necessarily outweigh the benefit of the high-pressure injection regarding fuel consumption. The total pump efficiency is a key factor for keeping the pump driving power below the effective power margin provided by the engine at high injection pressure. The total efficiency of the prototype pump is significantly higher than the total efficiency of the CP1. Therefore it enables to increase the injection pressure without causing an increase of the fuel consumption if the fuel mass flow is regulated. If the prototype pump always delivers the full maximum mass flow (no regulation), no fuel penalty should occur except at 6 bar IMEP and 80 MPa fuel pressure. But even at this engine operating point, the fictive driving power exceeds the effective power margin by 31 W only. Such an engine power variation and its corresponding fuel penalty is not significant.

A high total pump efficiency at high delivery pressure can be reached by maximizing the hydraulic, volumetric and mechanical efficiencies. The hydraulic efficiency can be increased by improving the design of the valves and fuel channels in the pump in order to reduce the pressure losses (especially at high flow rate). A smaller throttle gap between the pistons and the cylinders, a longer cylinder or reduced dead volumes in the work chambers enable to enhance the volumetric efficiency significantly. Low friction losses in the sliding systems lead to a higher mechanical efficiency. According to the results presented in previous chapter, the use of ceramic components in the sliding systems of a fuel-lubricated high-pressure pump for gasoline enable to reach a high mechanical efficiency. In contrast, the volumetric efficiency shows a high improvement margin, especially at low rotation speeds (high leakage level). Therefore, improving the volumetric efficiency is essential for the benefit/deficit balance at increased injection pressures.

The calculation provided in this section only concerns two operating points at mid-range engine speed. However the total pump efficiency is lower at low rotation speeds (higher friction losses and internal leakage). Consequently, ensuring a sufficient fuel supply at low rotation speeds may be achieved by a lower injection pressure (lower volumetric losses) or by an increased pump displacement. Reducing the injection pressure may reduce the benefit of high-pressure injection regarding pollutant emission. Increasing the pump displacement leads to a higher pump driving power and thus to a fuel penalty. As a result, it is essential to analyze the impact of the pump driving power on the engine performance over the entire operating map in order to make a compromise between the reduction of the pollutant emission and the reduction of the fuel consumption.

9 Summary

Spray-guided direct-injection shows great potential to decrease the fuel consumption of gasoline engines in comparison to the conventional port fuel injection or direct injection in homogeneous mode. However, this technology leads to new challenges since injecting the fuel directly into the combustion chamber may increase the soot emissions. This can be prevented by increasing the injection pressure significantly. The reduced droplet size resulting from the high fuel pressure leads to a faster vaporization and thus improves the mixture formation.

Conventional material combinations used in the sliding systems of fuel-lubricated high-pressure pumps wear severely at delivery pressures above 20 MPa, since the lubricity of gasoline is very low. To overcome these problems, advanced materials should be considered. Ceramics offer specific properties that may fit an application in fuel-lubricated high-pressure pumps. The investigations presented in this document aim at evaluating the performance of ceramic materials in the sliding systems, which are subjected to the highest tribological stresses. For that purpose, two prototype pumps have been designed as part of the collaborative research center SFB483 "High Performance Sliding and Friction Systems Based on Advanced Ceramics". A single-piston pump fitted with force transducers enables measurement of the friction coefficients in the cam/pusher and piston/cylinder systems at a delivery pressure of up to 50 MPa. A production-oriented 3-piston radial pump designed for a delivery pressure of up to 80 MPa allows measurement of the mechanical efficiency in order to evaluate the global friction losses with ceramic material pairs in the cam/pusher and piston/cylinder systems.

The investigations focus on the cam/pusher system, which is the most solicited in the high-pressure pump. Silicon carbide, sialon, and silicon nitride have been tested since these materials offer a high wear resistance, low friction coefficients and good chemical stability with gasoline and ethanol. Some cams made of silicon carbide have been textured with a laser in order to improve the lubrication of the sliding systems and prevent the abrasive action of particles released by wear. The material combinations have been investigated in stress cycles which reproduce real operating conditions of a high-pressure pump driven by the camshaft of a combustion engine. The pumps have been tested with isooctane, gasoline, ethanol, and gasoline-ethanol mixtures. After the experiment, the stressed surfaces of the sliding components have been analyzed on a scanning electron microscope in order to evaluate wear and identify which friction mechanisms occurred during operation. Complementary measurements have been performed with two commercially available pumps in order to better evaluate the improvements provided by the prototype pumps.

The results obtained with the single-piston prototype pump and isooctane as delivered fuel show that fine ground material pairs offer the best compromise between a fast run-in and stable performances. Self-mated silicon carbide and sialon with an α/β-phase ratio of 60/40 in the cam/pusher system lead to similar friction coefficients (in the range of 0.02 to 0.05). In contrast, the combination of silicon carbide with hardened bearing steel (AISI 52100) show increasing friction forces over time caused by adhesion effects. Texturing the cam prevents adhesion and leads to stable performance. However, the friction coefficient measured at low rotation speeds is significantly higher than with the self-mated ceramic pairs. Texturing the cam also improves the performance of self-mated silicon carbide. Gasoline provides a higher viscosity than isooctane and leads to lower friction coefficients. Using ethanol leads to even lower friction losses. Tribochemical reactions between ethanol and the investigated materials may contribute to a better lubrication as even a small quantity of ethanol mixed with gasoline results in very low friction levels. The low friction losses achieved with the best material combinations in the cam/pusher system reduce the tribological stresses in the piston/cylinder system. The friction losses measured with self-mated silicon carbide or with silicon carbide combined with AISI 52100 both correspond to approximately 1.5 % of the effective pump work. The surface analysis of the components tested in the cam/pusher system show that the surfaces have been smoothened but no significant material loss could be observed.

The investigations performed with the 3-piston prototype pump and gasoline confirm the results obtained with the single-piston pump. Since no significant difference could be observed between the two investigated material pairs in the piston/cylinder system of the single-piston pump, only silicon carbide combined with AISI 52100 was used in the piston/cylinder systems. A mechanical efficiency of better than 0.95 can be achieved with self-mated silicon carbide (textured or not) or with self-mated sialon (with an α/β-phase ratio of 60/40) in the cam/pusher systems. Texturing the cam of the self-mated silicon carbide pairs extends the high-efficiency range in the pump operating map in comparison to the untextured pair. Textured silicon carbide cams combined with pushers made of AISI 52100 show good performance with a mechanical efficiency of up to 0.95 and without any adhesion mechanism observed in the sliding systems. Even if it provides a higher hardness, increasing the ratio of α-phase in the parts made of sialon causes higher friction losses. Moreover, micro-cracks were observed on the surface of the sialon components with the α/β-phase ratio of 90/10, thus indicating a mechanism of surface fatigue. In contrast to the material pairs based on silicon carbide, both investigated sialon pairs lead to noise emissions and cyclic variations of the pump driving torque. Using a sialon-silicon carbide composite prevents these phenomena but provides a lower mechanical efficiency in comparison to self-mated silicon carbide or self-mated sialon with an α/β-phase ratio of 60/40. Self-mated silicon nitride does not offer satisfactory performance regarding friction losses in the sliding systems and leads to high noise emissions. Combining silicon nitride with hardened bearing steel provides better performances than the self-mated pair but the pump mechanical efficiency is below 0.85 over the entire investigated operating map. In contrast, silicon nitride containing carbon-nanotubes combined with hardened bearing steel offers very good performance regarding mechanical efficiency, noise emission, and cyclic variations of the pump driving torque. The significant improvement provided by the carbon-nanotubes added to the silicon nitride matrix may be caused by the deformation capability of the carbon-nanotubes, which enables a better distribution of the local stresses on the surface of the bodies. In addition, the analysis of the stressed surfaces suggests tribochemical reactions which may contribute to an improved lubrication of the sliding system.

The experiments performed with production pumps have shown decreasing mechanical and volumetric efficiencies over time. These effects are linked to the high friction losses which cause a reduced piston motion and negatively affect the suction stroke. The surfaces of the sliding components show macroscopic scratches and disappearing coatings after a few hours of operation only, thus indicating high friction and wear levels. The use of ceramic components in the sliding systems of fuel-lubricated high-pressure pumps for gasoline prevents these issues from occurring at high delivery pressure levels.

Increasing the delivery pressure of high-pressure gasoline pumps induce some further challenges such as providing a high volumetric efficiency. The volumetric losses due to the increased internal leakage and the longer re-expansion of the fuel in the dead volumes affect the total pump efficiency. This deficit must be balanced by a larger displacement. Since the increase of the delivery pressure and the reduction of the volumetric efficiency result in an increase of the pump driving torque, the impact of injection pressure levels well above 20 MPa may cause a higher fuel consumption by the combustion engine. Exemplary calculations have shown that no fuel penalty should occur with the total efficiency level of the 3-piston prototype pump developed as part of this work. However, these calculations were performed at two engine operating points only. Further investigations of the entire propulsion system (including the combustion engine and the mechanically driven high-pressure pump) are essential to evaluate the real impact of higher injection pressures on the engine performance over the whole engine operating map. Such experiments would allow developers to find a compromise between the reduction of the pollutant emission, the potential fuel penalty and the reliable operation of the high-pressure pump over time.

Abbreviations

BDC Bottom Dead Centre

CA Crank Angle

CNT Carbone NanoTube

CO$_2$ Carbon Dioxide

EDX Energy Dispersive X-ray spectroscopy

GDI Gasoline Direct Injection

IAM-KM Institut für Angewandte Materialien - Keramik im Maschinenbau

IAM-WK Institut für Angewandte Materialien - Werkstoffkunde

IMEP Indicated Mean Effective Pressure

IPCC Intergovernmental Panel on Climate Change

ISFC Indicated Specific Fuel Consumption

KIT Karlsruhe Institute of Technology

PFI Port Fuel Injection

RPM Revolutions Per Minute

RON Research Octane Number

SEM Scanning Electron Microscope

Si$_3$N$_4$ Silicon Nitride

SiC Silicon Carbide

SMD Sauter Mean Diameter

SSiC Sintered Silicon Carbide

TDC Top Dead Centre

Bibliography

[1] *BWK das Energie-Fachmagazin*, volume 4. Springer VDI Verlag, 2012.

[2] S. M. Abo-Naf, U. Dulias, J. Schneider, K.-H. Zum Gahr, S. Holzer, and M. J. Hoffmann. Mechanical and tribological properties of nd- and yb-sialon. *Journal of Materials Processing Technology*, 183:264–272, 2007.

[3] W. Anderson, J. Yang, D. D. Brehob, J. K. Vallance, and R. M. Whiteaker. Understanding the thermodynamics of direct injection spark ignition (disi) combustion systems: An analytical and experimental investigation. *SAE Technical Paper n° 962018*, 1996.

[4] Y. Iwamoto andK. Noma, O. Nakayama, T. Yamauchi, and H. Ando. Development of gasoline direct injection engine. *SAE Technical Paper n° 970541*, 1997.

[5] European Investment Bank. Eib supports new nissan battery plant and electric car production in sunderland. *http://www.eib.org*, 2011.

[6] European Investment Bank. Supporting electric vehicle development with bolloré. *http://www.eib.org*, 2011.

[7] European Investment Bank. Electric vehicles: Eib lends eur 180 million to renault group. *http://www.eib.org*, 2012.

[8] H.-J. Bargel and G. Schulze. *Werkstoffkunde*. VDI Verlag, 1994.

[9] R. Van Basshuysen. *Ottomotor mit Direkteinspritzung*. Springer, 2007.

[10] R. Bauder, J. Helbig, H. Marckwardt, and H. Genc. Der neue 3,0-l-tdi-biturbomotor von audi. *MTZ Motortechnische Zeitschrift, Ausgabe 1*, 2012.

[11] C. Baumgarten. *Mixture formation in internal combustion engine*. Springer, 2006.

[12] W. Bernhart. Electromobility - the only way forward? *19. Aachener Kolloquium Fahrzeug- und Motorentechnik*, 2010.

[13] J. Böhme, W. Hatz, A. Eiser, R. Dornhöfer, W. Ehret, and R. Wurms. Der neue r4 1,8l t-fsi-motor von audi. *15. Aachener Kolloquium Fahrzeug- und Motorentechnik*, pages 107–144, 2006.

[14] BP. *BP Statistical Review of World Energy*. http://www.bp.com, 2012.

[15] S. Buri. *Untersuchungen des Potenzials von Einspritzdrücken bis 1000 bar in einem Ottomotor mit Direkteinspritzung und strahlgeführtem Brennverfahren.* Logos Verlag, 2011.

[16] S. Buri, S. Busch, H. Kubach, and U. Spicher. High injection pressures at the upper load limit of stratified operation in a disi engine. *SAE Technical Paper n° 2009-01-2657*, 2009.

[17] S. Buri, C. Dahnz, H. Kubach, and U. Spicher. Reduction of soot emissions by increasing injection pressure up to 1000 bar in a disi engine in stratified operation. *9th International Symposium on Combustion Diagnostics*, 2010.

[18] E. J. Candès and M. B. Wakin. An introduction to compressive sampling. *IEEE Signal Processing Magazine*, 2008.

[19] C. Barry Carter and M. Grant Norton. *Ceramic Materials, Science and Engineering.* Springer, 2007.

[20] H. Czichos and K.-H. Habig. *Tribologie Handbuch.* Vieweg, 2010.

[21] A. Dasari, Z.-Z. Yu, and Y.-W. Mai. Fundamental aspects and recent progress on wear/scratch damage in polymer nanocomposites. *Materials Science and Engineering Reports*, 63:31–80, 2009.

[22] Delphi. Delphi premium gdi high pressure fuel pump. http://delphi.com/shared/pdf/ppd/pwrtrn/premium-gdi-high-pressure-fuel-pump.pdf, 2012.

[23] Y. Duchaussoy, B. Covin, Y. Boccadoro, O. Meurisse, J. P. Mercier, and D. Levasseur. Renault energy tce 115 - the new renault 1.2 gdi turbocharged engine. *20th Aachen Colloquium Automobile and Engine Technology*, 2011.

[24] H. Eichlseder, M. Klüting, and W. F. Piock. *Grundlagen und Technologien des Ottomotors.* Springer, 2008.

[25] T. Ekström and M. Nygren. Sialon ceramics. *Journal of the American Ceramics Society*, 75:259–276, 1992.

[26] C. Enderle, M. Mürwald, G. Tiefenbacher, G. Karl, and P. Lautenschutz. *Neue Vierzylinder-Ottomotoren von Mercedes-Benz mit Kompressoraufladung.* MTZ Motortechnische Zeitschrift, Ausgabe 7, 2002.

[27] D. Gagliardi and D. Klein. Strategies for the worldwide race to co2 reduction. *Stuttgart Symposium*, 2010.

[28] N. Gebhardt. *Pumpen und Motoren.* Springer, 2008.

[29] B. Geringer and K. Tober. Zukünftige mobilität: Elektromobilität als lösung? Technical report, Österreichischer Verein für Kraftfahrzeugtechnik, 2010.

[30] Robert Bosch GmbH. *Dieselmotor-Management*. Vieweg, 2004.

[31] Robert Bosch GmbH. *Ottomotor-Management*. Vieweg, 2005.

[32] J. Gonzalez-Julian, J. Schneider, P. Miranzo, M. I. Osendi, and M. Belmonte. Enhanced tribological performance of silicon nitride-based materials by adding carbon nanotubes. *JACE, Manuscript n°04391*, 2011.

[33] I. G. Goryacheva. *Contact Mechanics in Tribology*. Kluwer Academic, 1998.

[34] K.H. Grote and E. K. Antonsson. *Handbook of Mechanical Engineering*. Springer Berlin Heidelberg, 2008.

[35] J. Harada, T. Tomita, H. Mizuno, Z. Mashiki, and Y. Ito. Development of direct injection gasoline engine. *SAE Technical Paper n° 970540*, 1997.

[36] T. Heiduk, M. Kuhn, M. Stichlmeir, and F. Unselt. *Der neue 1,8-L-TFSI-Motor von Audi*. MTZ Motortechnische Zeitschrift, 2008.

[37] J. P. Häntsche. *Entwicklung und experimentelle Untersuchung einer Hochdruckpumpe für Ottokraftstoff basierend auf ingenieurkeramischen Gleitsystemen*. Logos, 2009.

[38] J. P. Häntsche, G. Krause, A. Velji, and U. Spicher. High pressure fuel pump for gasoline direct injection based on ceramic components. *SAE Technical Paper n° 2005-01-2103*, 2005.

[39] H. Izumida, T. Nishioka, A. Yamakawa, and M. Yamagiwa. A study of the effects of ceramic valve train parts on reduction of engine friction. *SAE Technical Paper n° 970003*, 1997.

[40] K. H. Jack. Sialons and related nitrogen ceramics. *Journal of materials science*, 11:1135–1158, 1976.

[41] R. W. Jorach, P. Bercher, G. Meissonnier, and N. Milovanovic. Common-railsystem von delphi mit magnetventilen und einkolben-hochdruckpumpe. *MTZ Motortechnische Zeitschrift, Ausgabe 3*, 2011.

[42] Informationszentrum Technische Keramik. *Brevier Technische Keramik*. Fahner Verlag, 1999.

[43] F. Kessler, G. Kiesgen, J. Schopp, and M. Bollig. *Die neue Vierzylinder-Motorenbaureihe aus der BMW/PSA-Kooperation*. MTZ Motortechnische Zeitschrift, Ausgabe 7, 2007.

[44] A. Kneifel. *Hochdruckeinspritzung als Möglichkeit zur Kraftstoffverbrauchs- und Abgasemissionsreduzierung bei einem Ottomotor mit strahlgeführtem Brennverfahren*. PhD thesis, Karlsruher Institut für Technologie (KIT), 2008.

[45] Kraftfahrt-Bundesamt. *Emissionen, Kraftstoffe - Deutschland und seine Länder am 1. Januar 2012*. http://www.kba.de, 2012.

[46] R. Krebs, R. Szengel, H. Middendorf, H. Sperling, W. Siebert, J. Theobald, and K. Michels. *Neuer Ottomotor mit Direkteinspritzung und Doppelaufladung von Volkswagen - Teil 2: Thermodynamik.* MTZ Motortechnische Zeitschrift, Ausgabe 12, 2005.

[47] K. H. Küttner. *Kolbenmaschinen.* Vieweg+Teubner, 2009.

[48] O. Kunde, J. Hansen, T. Zenner, P. Kapus, C. Obst, and B. Queenan. The new 2.0 scti ecoboost gasoline engine from ford. *19. Aachener Kolloquium Fahrzeug- und Motorentechnik,* 2010.

[49] B Lechner, K. Kiesgen, J. Kriese, and J. Schopp. *Der neue Mini-Motor mit Twin-Power-Turbo.* MTZ Motortechnische Zeitschrift, Ausgabe 7, 2010.

[50] A. H. Lefebvre. *Atomization and Sprays.* Taylor and Francis, 1989.

[51] G. Leuschner. *Kleines Pumpen-Handbuch für Chemie und Technik.* Verlag Chemie GmbH, 1967.

[52] F. Mathieu, M. Reddemann, D. Martin, and R. Kneer. Experimental investigation of fuel influence on atomization and spray propagation using an outwardly opening gdi-injector. *SAE Technical Paper n° 2010-01-2275,* 2010.

[53] G. P. Merker and C. Schwarz. *Grundlagen Verbrennungsmotoren.* Vieweg+Teubner, 2011.

[54] H. Middendorf, J. Theobald, L. Lang, and K. Hartel. *Der 1,4-L-TSI-Ottomotor mit Zylinderabschaltung.* MTZ Motortechnische Zeitschrift, Ausgabe 3, 2012.

[55] N. Mitroglou, J. M. Nouri, M. Gavaises, and C. Arcoumanis. Spray characteristics of a multi-hole injector for direct-injection gasoline engines. *International Journal of Engine Research,* 2006.

[56] N. Mitroglou, J. M. Nouri, Y. Yan, M. Gavaises, and C. Arcoumanis. Spray structure generated by multi-hole injectors for gasoline direct-injection engines. *SAE Technical Paper n° 2007-01-1417,* 2007.

[57] M. Mittwollen and H. Pfeiffer. *Literaturrecherche über die tribologischen Eigenschaften von Keramik.* VDI Verlag, 1988.

[58] F. Mogge and M. Berret. Automotive landscape 2025. *Zeitschrift für die gesamte Wertschöpfungskette,* 1, 2012.

[59] A. Nauwerck. *Untersuchung der Gemischbildung in Ottomotoren mit Direkteinspritzung bei strahlgeführtem Brennverfahren.* PhD thesis, Karlsruhe Institute of Technology (KIT), 2006.

[60] A. Nauwerck, J.Pfeil, A.Velji, and U. Spicher. A basic experimental study of gasoline direct injection at significantly high injection pressures. *SAE Technical Paper n° 2005-01-0098,* 2005.

[61] B. Nesbitt. *Handbook of pumps and pumping.* Elsevier, 2006.

[62] Intergovernmental Panel on Climate Change. *Climate Change 2007: Synthesis Report.* http://www.ipcc.ch, 2007.

[63] J. Pape, A. Kneifel, O. Tremmel, A. Velji, and U. Spicher. 'bridging the ring gap' with pressure and holes - the approach to an affordable injector for gasoline direct injection systems of the future. *SIA Spark Ignition International Conference,* 2007.

[64] European Parliament. *Commission regulation (EU) 459/2012 of 29 May 2012 amending Regulation (EC) No 715/2007 of the European Parliament and of the Council and Commission Regulation (EC) No 692/2008 as regards emissions from light passenger and commercial vehicles (Euro 6).* http://europa.eu, 2012.

[65] W. Piock, G. Hoffmann, A. Berndorfer, P. Salemi, and P. Fusshoeller. Strategies toward meeting future particulate matter emission requirements for si engines. *SAE Technical Paper n° 2011-01-1212,* 2011.

[66] W. Pohlenz. *Grundlagen für Pumpen.* VEB Verlag Technik Berlin, 1975.

[67] V. L. Popov. *Kontaktmechanik und Reibung.* Springer Berlin Heidelberg, 2010.

[68] K. Reif. *Moderne Diesel-Einspritzsysteme.* Vieweg+Teubner, 2010.

[69] M. Riva. *Entwicklung und Charakterisierung von Sialon-Keramiken und Sialon-SiC-Verbunden für den Einsatz in tribologisch hochbeanspruchten Gleitsystemen.* KIT Scientific Publishing, 2011.

[70] M. Riva, M. J. Hoffmann, R. Oberacker, and T. Fett. Subcritical crack growth in sialon ceramics from a modified static lifetime test including multiple use of survivals. *Journal of materials science,* 43:402–405, 2007.

[71] H. Salmang and H. Scholze. *Keramik.* Springer, 2007.

[72] T. W. Scharf, A. Neira, J. Y. Hwang, J. Tiley, and R. Banerjee. Self-lubricating carbon nanotube reinforced nickel matrix composites. *Journal of applied physics,* 106, 2009.

[73] F. Schumann, S. Buri, C. Dahnz, H. Kubach, and U. Spicher. Emissionen bei strahlgeführter benzindirekteinspritzung mit einspritzdrücken bis 1000 bar. In *Direkteinspritzung im Ottomotor 8.* Expert, 2011.

[74] F. Schumann, S. Buri, H. Kubach, and U. Spicher. Investigation of particulate emissions from a disi engine with injection pressures up to 1000 bar. *19. Aachener Kolloquium Fahrzeug- und Motorentechnik,* 2010.

[75] F. Schumann, H. Kubach, and U. Spicher. The influence of injection pressures of up to 800 bar on catalyst heating operation in gasoline direct injection engines. *8th International Conference on Modeling and Diagnostics for Advanced Engine Systems*, 2012.

[76] C. Schwarz, S. Missy, H. Steyer, B. Durst, E. Schünemann, W. Kern, and A. Witt. *Die neuen Vier- und Sechszylinder-Ottomotoren von BMW mit Schichtbrennverfahren*. MTZ Motortechnische Zeitschrift, Ausgabe 5, 2007.

[77] M. Sens, J. Maass, S. Wirths, and R. Marohn. Effects of highly heated fuel and/or high-pressure injection on spray formation in gasoline direct-injection engines. *SIA Spark Ignition International Conference*, 2011.

[78] M. Skogsberg, P. Dahlander, and I. Denbratt. Spray shape and atomization quality of an outward-opening piezo gasoline di injector. *SAE Technical Paper n° 2007-01-1409*, 2007.

[79] K. Sommer, R. Heinz, and J. Schöfer. *Verschleiß metallischer Werkstoffe Erscheinungsformen sicher beurteilen*. Vieweg+Teubner, 2010.

[80] T. D. Spegar, S.-I. Chang, S. Das, E. Norkin, and R. Lucas. An analytical and experimental study of a high pressure single piston pump for gasoline direct injection (gdi) engine applications. *SAE Technical Paper n° 2009-01-1504*, 2009.

[81] U. Spicher. Is there a future for combustion engines? *10. Internation Symposium on Combustion Diagnostics*, 2012.

[82] U. Spicher, J. Reissing, J.M. Kech, and J. Gindele. Gasoline direct injection (gdi) engines - development potentialities. *SAE Technical Paper n° 1999-01-2938*, 1999.

[83] T.A. Stolarski. *Tribology in Machine Design*. Elsevier, 1990.

[84] R. Szengel, H. Middendorf, E. Pott, J. Theobald, T. Etzrodt, and R. Krebs. The tsi with 90 kw - the expansion of the volkswagen family of fuelefficient gasoline engines. *28. Internationales Wiener Motorensymposium*, 2007.

[85] H.-D. Tietz. *Technische Keramik*. VDI Verlag, 1994.

[86] E. Truckenbrodt. *Fluidmechanik*. Springer, 2008.

[87] H. Tschöke and H. Hölz. *Verdrängerpumpen*. Springer, 2011.

[88] R. van Basshuysen and F. Schäfer. *Lexikon Motorentechnik*. Vieweg, 2004.

[89] G. Vetter. *Pumpen*. Vulkan-Verlag Essen, 1992.

[90] A. Waltner, P. Lückert, G. Doll, and R. Kemmler. *Der neue 3,5-l-V6-Ottomotor mit Direkteinspritzung von Mercedes-Benz*. MTZ Motortechnische Zeitschrift, Ausgabe 9, 2010.

[91] A. Waltner, P. Lückert, G. Doll, and R. Kemmler. *Der neue 4,6-l-V8-Ottomotor von Mercedes-Benz.* MTZ Motortechnische Zeitschrift, Ausgabe 10, 2010.

[92] H. Watter. *Hydraulik und Pneumatik.* Vieweg+Teubner, 2008.

[93] D. P. Wei, H. A. Spikes, and S. Korcek. The lubricity of gasoline. In *Tribology Transactions.* Taylor & Francis, 1999.

[94] A. Welter, H. Unger, U. Hoyer, T. Brüner, and W. Kiefer. The new turbocharged bmw six cylinder inline petrol engine. *15. Aachener Kolloquium Fahrzeug- und Motorentechnik*, pages 89–106, 2006.

[95] J. Willand, K. Schintzel, and H. Hoffmeyer. *Das Potenzial aufgeladener Ottomotoren mit Direkteinspritzung.* MTZ Motortechnische Zeitschrift, Ausgabe 2, 2009.

[96] E. Winklhofer and G. Fraidl. Developing clean gdi systems to meet euro 6 targets. *10th International Symposium on Combustion Diagnostics*, 2012.

[97] M. Winterkom, L Spiegel, P. Bohne, and G. Söhlke. *Der Lupo FSI von Volkswagen - So sparsam ist sportlich.* MTZ Motortechnische Zeitschrift, Ausgabe 11, 2000.

[98] M. Wöppermann. *Einfluss einer Wirkflächentexturierung auf das tribologische Verhalten von Stahl/Keramik-Paarungen unter reversierender mediengeschmierter Gleitbeanspruchung.* PhD thesis, Karlsruher Institut für Technologie (KIT), 2011.